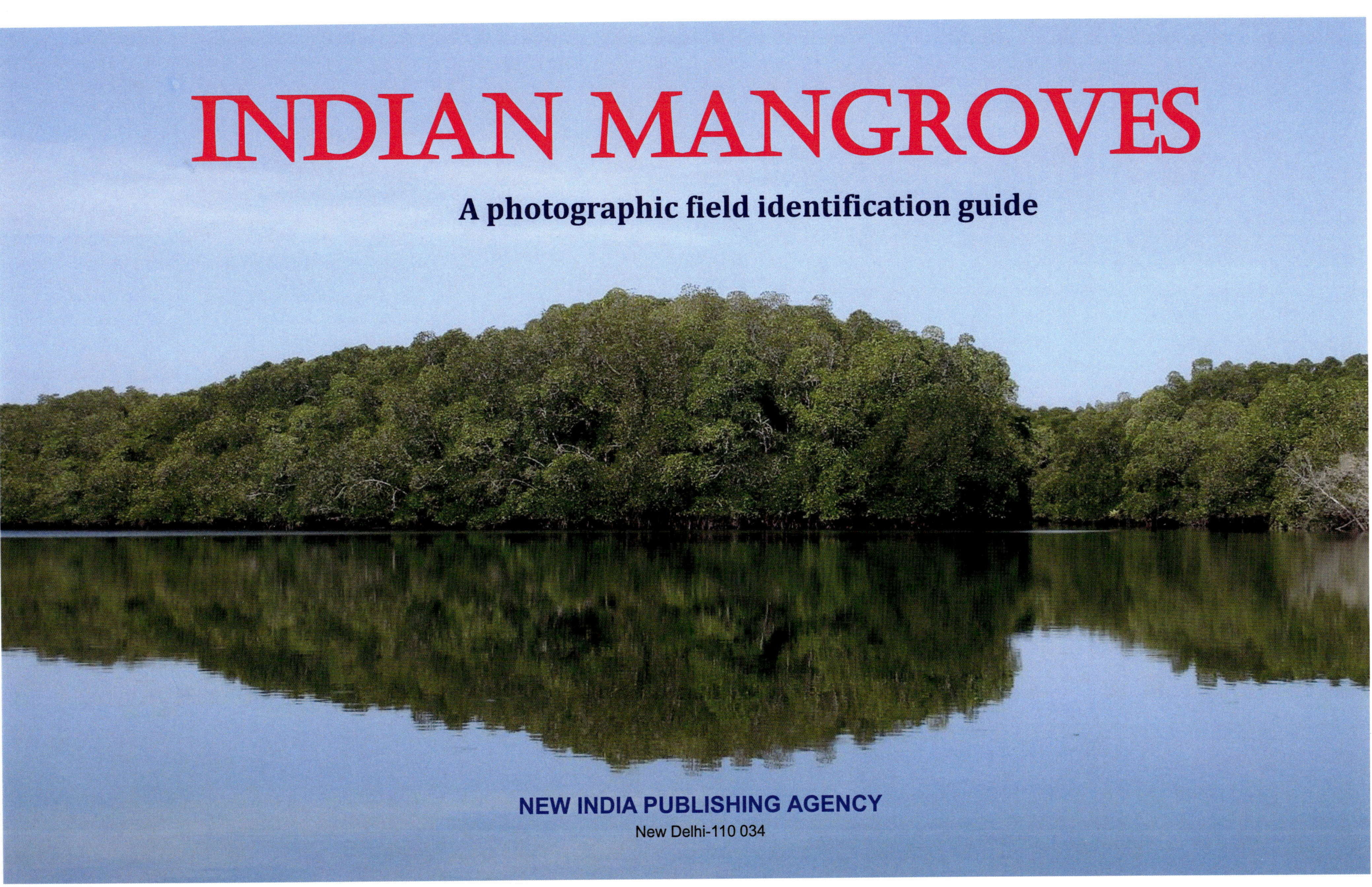

INDIAN MANGROVES

A photographic field identification guide

NEW INDIA PUBLISHING AGENCY
New Delhi-110 034

INDIAN MANGROVES
A photographic field identification guide
P. Ragavan
K. Kathiresan
T.S. Rana
Alok Saxena
P.M. Mohan
R.S.C. Jayaraj
K. Ravichandran
T. Mageswaran

New India Publishing Agency

101, Vikas Surya Plaza, CU Block, LSC Market
Pitam Pura, New Delhi – 110 034, India
Email: info@nipabooks.com
Web: www.nipabooks.com

For customer assistance, please contact
Phone: +91 -11-27 34 17 17 Fax:+91 -11-27 34 16 16
E-Mail: feedbacks@nipabooks.com

ISBN : 978-93-90175-99-4

Composed and Designed by NIPA

INDIAN MANGROVES

A photographic field identification guide

Authors

P. Ragavan
K. Kathiresan
T.S. Rana
Alok Saxena
P.M. Mohan
R.S.C. Jayaraj
K. Ravichandran
T. Mageswaran

Population of *Avicennia marina* in the Andaman Islands

Foreword

"Mangroves" – the word conjures up so many magnificent images and exciting thoughts of a very special marine and coastal ecosystem - a watery place that dries with daily tides, of specially adapted plants, of complex breathing roots, of colourful exotic flowers, of fruits in many shapes and sizes, of buoyant seedling propagules, of salt crusted leaves, of teaming small animals, of succulent shellfish, and much more!

All this wonderful natural diversity almost secretly surrounds the Indian subcontinent but it is rarely fully appreciated! This extraordinary book opens up this largely hidden place with its informative descriptions of not only the plants themselves, but also their many ecosystem benefits given freely to human kind - like protecting our shorelines, like supplying beautiful fish and marine life to eat, like cleaning coastal waters as kidneys do for blood, like capturing and storing harmful atmospheric carbon, and with a seemingly endless supply of firewood and honey. All we have to do is look after these vulnerable places!

"Indian Mangroves" – much more than a guide! This carefully-researched and factual account of the mangroves of India expertly describes each plant with great care and thought, providing a rich and valuable resource of reliable facts and informative colour photographs to help you identify each and every species and type of mangrove plant known around the country. With its' publication, this impressive work has become the magnum opus of authoritative botanical literature on India's' mangrove plants – a national treasure and a lasting reference for decades to come.

In closing, this rare collaboration of leading authorities on Indian mangroves are to be roundly commended and congratulated for bringing this impressive account to the people with such treasured thoughts and words about our magnificent mangroves.

****

Professor Norman C Duke, MSc, PhD
Mangrove Researcher, James Cook University
Director CEO, Mangrove Watch
Member, IUCN Mangrove Specialist Group
Vice President, International Society of Mangrove Ecosystems
Chief Facilitator, Australian Mangrove and Saltmarsh Network
Life Member, Mangrove Society of India
7 May 2020

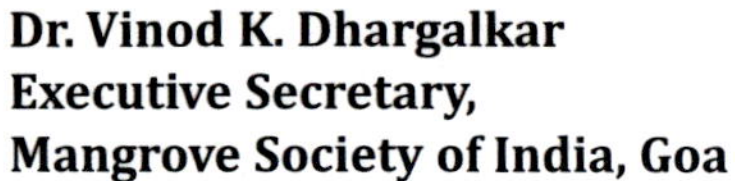

Dr. Vinod K. Dhargalkar
Executive Secretary,
Mangrove Society of India, Goa

Foreword

Mangrove ecosystem plays a significant role in protecting life and livelihood of the coastal communities of the world. Mangroves are widely distributed with continuous as well as fragmented patches along tropics and sub-tropical region. They are diverse group of woody trees, palms, shrubs, and ferns that share a common ability to live in waterlogged saline soils subjected to regular flooding. These halophytic, highly specialized plants have developed unusual morphological, anatomical and physiological adaptations to sustain the unique environmental conditions.

In the past few decades, mangrove forests have been largely degraded and destroyed during the process of developmental activities, pollution and climate change. The negative environmental and socio-economic impacts on mangrove ecosystems have resulted in many government and non-government agencies, along with NGO's and civil societies to undertake mangrove conservation, rehabilitation and management programmes.

During such exercise, researchers and working staff have faced continual difficulties in identifying plant species in the field. Although, there are number of mangrove manuals and field guidebooks, none of these, sufficiently cover species that include not only grasses, herbs and ferns, but also trees, shrubs and climbers.

It gives me great pleasure and honour to introduce this unique publication "Indian Mangroves - A photographic field identification guide to India's mangrove plants" to those who study mangroves, especially the younger generation, to learn about mangrove forests in India. The author has sighted about 46 species that forms integral parts of this coastal ecosystem, were identified and documented with photographs and brief descriptions. The book further describes the structure and function of mangroves, their value and need for sustainable conservation.

We should all acknowledge with thanks the work done by the main authors headed by Dr. P. Ragavan and other expert co-authors for painstaking efforts, in bringing out this interesting, informative and beautifully illustrated publication.

I am confident that this publication will be a useful tool for mangrove forest managers, foresters, coastal resource managers, scientists, students and those interested lay persons, not only in India but also in many other countries where mangroves grow.

Dr. Vinod K. Dhargalkar

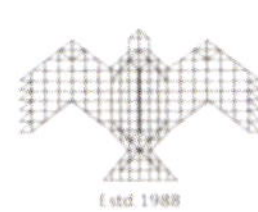

Dr. Shailesh Nayak
Director

NATIONAL INSTITUTE OF ADVANCED STUDIES
Indian Institute of Science Campus, Bangalore - 560 012, INDIA

Tele: +91-80-2360 1969 (O) Fax: +91-80-2218 5028
Email: director@nias.res.in ; shailesh@nias.res.in

Foreword

"Mangroves"- a vital ecosystem of the tropical and sub-tropical coast of the world, are nature's gift for the welfare of millions of coastal peoples. Mangroves grow in clayey and silty intertidal areas, deltaic and estuarine coasts and backwaters/sheltered regions. Mangrove provide habitat for variety of organisms and spawning and nursery grounds of fish as well as protect coast from erosion. Many areas in India have lost mangroves over the past 50 years and have affected the environment, the climate, and humanity, through diminished benefits such as carbon storage, coastal protection and fish protection. Thus mangrove conservation should be prioritized globally.

This amazing book showcases the beauty and diversity of mangrove plants in India. Detailed description, photographs and distribution of each and every mangrove species known around the country help to strengthen the existing knowledge base for better management. Inadequate understanding of the mangrove plants is often identified as causes for failure many restoration/rehabilitation initiatives. So, it is anticipated that this field guide will be used for identification that will promote greater protection, increased propagation, as well as better appreciation of these species and the forests in which they occur. I am pretty sure that this publication will be an important resource for community, land managers, forest staff and visitors alike and become a standard reference of Indian mangroves for decades to come.

In closing, I appreciate the efforts of authors for bringing this impressive work and I wish them to achieve success in all their future endeavours.

Shailesh Nayak

About the Authors

Dr. P. Ragavan is a National Post-Doctoral Research Fellow at CSIR-National Botanical Research Institute, Lucknow. He was awarded a Ph.D degree in Ocean Studies and Marine Biology from Pondicherry University. He has more than ten years of research experiences on his area of interest viz., taxonomy and ecology of mangroves and seagrasses. More recent research has concentrated on molecular taxonomy and carbon storage potential of Indian mangroves.

Prof. Dr. K. Kathiresan has spent practically most of his career on mangroves. He is Doctor of Science (D.Sc.) in Marine Biology and is currently an honorary professor in Annamalai University. He has about 500 publications and 10 books. He has made significant contributions to mangroves for their biodiversity, ecology, conservation and management. He has conducted many national and international training programmes on mangroves. He is a recipient of many national and international awards and distinctions. He studied the mangroves in India and 16 other countries. He served as a mangrove expert in top level committees.

Dr. T. S. Rana is presently working as a Chief Scientist, Area Coordinator & Head in Plant Diversity, Systematics and Herbarium Division at CSIR-National Botanical Research institute, Lucknow (India). He has over three decade's research experience in Plant Taxonomy, Molecular Systematics, DNA Fingerprinting and Bio-prospecting of Medicinal Plants. Dr. Rana has published four books and over 110 research papers in peer reviewed National and International journals. He has supervised/guided eight PhD students and mentored over half a dozen post doctorate fellows. He is a life member and fellow of various renowned scientific bodies. Dr. Rana was bestowed with prestigious BOYSCAST Fellowship (2001-2002) by the Department of Science and Technology, Ministry of Science and Technology, Government of India, New Delhi for his outstanding contributions in Plant Sciences.

Dr. Alok Saxena is retired Principal Chief Conservator of Forests, Andaman & Nicobar Administration. Currently working as Senior Consultant at Wildlife Institute of India, Dehradun, India. He was awarded a PhD in Biochemistry from the Central Drug Research Institute, Lucknow in 1983. He contributed significantly to State of Forest Reports (SFRs) in 1993, 1995, 1997, 2001 and 2003. He has many research publications in a wide variety of fields, such as biochemistry, forest fire, climate change, mangroves, coral reefs, and assessment of natural resources.

Dr. P.M. Mohan is Professor of Marine Biology at Pondicherry University. He was awarded a PhD in Marine Sciences from Cochin University of Science and Technology, Kerala, India. He was the founder and head of the Department of Ocean Studies and Marine Biology, Pondicherry University, Andaman and Nicobar Islands in 2002. His expertise is on the monitoring of critical habitats, namely, mangroves, coral reefs and seagrass meadows.

Dr. R.S.C. Jayaraj is presently the Director of Rain Forest Research Institute, Jorhat, Assam. He was awarded Ph.D in Forest Ecology and Environment. He joined the Indian Forest Service in 1987. He has served as Silviculturist in the Andaman and Nicobar Islands and at the Institute of Forest Genetics and Tree Breeding, Coimbatore, India. He was earlier the Director of State Forest Research Institute, Itanagar, Arunachal Pradesh. His expertise is on breeding of Eucalyptus and Casuarina, besides work on mangroves and ecophysiology.

Dr. K. Ravichandran is the Chief Conservator of Forests in the Department of Environment and Forests, Union Territory of Dadra & Nagar Haveli. He was awarded a PhD in forestry from the Forest Research Institute, Dehradun, India. He joined the Indian Forest Service in 1996. His areas of interest include tropical forest management, coastal and marine ecosystems and conservation and management of critical habitats, namely, mangroves, coral reefs and seagrass meadows.

Dr.T. Mageswaran is working as a scientist in the National Centre for Sustainable Coastal Management (NCSCM), Tamil Nadu, India. He was awarded a PhD in Geology from University of Madras, Chennai, India. His expertise is on coastal zone management, tsunami hazard, disaster management and spatial mapping etc. He has more than ten years of experience in the field of Remote Sensing and GIS.

Preface

India is one of the 17 mega biodiversity countries in the world. However, taxonomic inventory of India biota is far from complete. The goal of this publication is to identify and describe the mangrove plant species of India. Towards this objectives, the taxonomical identity and distribution of mangrove species found in the country is revised and described. During field investigations, each species of mangrove and associated plants were identified and sampled. Description and high-resolution image material for each species is provided, which allows easy identification of individual species. This information is needed for coastal management planning and policy development, especially in relation to shoreline rehabilitation and expansion of appropriate shoreline livelihood projects in the face of climate change and sea level rise. Furthermore, global distribution of each mangrove species and its conservation status is provided. Voucher specimens of all the mangrove species reported in the text have been deposited at CSIR-National Botanical Research Institute, Lucknow (LWG) as well in Annamalai University and regional centres of Botanical Survey of India as reference collections for the country.

Dr. P. Ragavan

Extensively developed stilt roots of *Rhizophora apiculata*

Acknowledgements

I take this opportunity to express my deep sense of gratitude towards all the researchers who have worked on Indian mangroves. Here I would like to thank Dr. J.C. Dagar, Prof. A.G. Untawale, Dr. T.G. Jagtap, Dr. K.R. Naskar, Dr. R.N. Mandal, Dr. L.K. Banerjee, Dr. T.A. Rao and Prof. Leela Bhosale for their inspirational work on mangroves of India. Indian mangroves, without their footprints are meaningless.

I have great pleasure in expressing my gratitude to Professor N.C. Duke, James Cook University, Australia; Prof. Peter Saenger, Southern Cross University, Australia; Shri Tarun Coomar, PCCF, Andaman and Nicobar Administration; Shri D.M. Shukla PCCF, Andaman and Nicobar Administration; Dr. C. Murugan, Botanical Survey of India; Prof. R. Ramesh, Director, NCSCM, Chennai; Dr. P. Krishnan, Principal Scientist, ICAR-NAARM, Hyderabad; Prof. S.K. Barik, Director, CSIR-NBRI, Lucknow; Prof. L. Kannan, Emeritus Professor, Annamalai University; Dr. K. Sivakumar, Associate Professor, Annamalai University; Dr. Sanjeev Kumar, Associate Professor, Physical Research Laboratory, Ahmedabad; Dr. B. Nagarajan, Scientist G, IFGTB, Coimbatore; and, Shri N. Vasudevan, APCCF, Maharashtra Forest Department, for their support, guidance and helpful suggestions.

I owe my thanks to Forest Department staff of Andaman & Nicobar Islands, West Bengal, Odisha, Andhra Pradesh, Maharashtra, Tamil Nadu, Kerala, Gujarat, Karnataka and Goa, and staff of Pondicherry University, Puducherry, Institute of Forest Genetics and Tree Breeding, Coimbatore, ICAR-Central Island Agricultural Research Institute, Port Blair, National Centre for Sustainable Coastal Management, Chennai, CSIR-National Botanical Research Institute, Lucknow, and Physical Research Laboratory, Ahmedabad for support and cooperation in various ways.

I extend my thanks to Mr. Muktipada Panda, Assistant Professor, Regional Institute of Education, Bhubaneswar, National Council of Educational Research and Training, Government of India and Dr. S. K. Dubey for their support in the field at Odisha and Sundarbans respectively.

Special thanks to Science and Engineering Research Board, Department of Science and Technology, New Delhi and Department of Space, Govt. of India for funding my post-doctoral research at CSIR-NBRI, Lucknow (2017-2019) and PRL, Ahmedabad (2020-2022) respectively.

Finally, but not the end, I am always thankful to my family, friends, well-wishers, critics and all others who helped me directly or indirectly in the completion of the book.

Dr. P. Ragavan

Growth of *Acrostichum aureum* in association with *Rhizophora* spp.,

Contents

Population of *Nypa fruticans* in the Andaman Islands

Part 1: Mangroves an Overview

Mangroves Definition

A mangrove is tree, shrub, palm or ground fern, generally exceeding one and a half meter in height that normally grows above mean sea level in the intertidal zone of marine coastal environment and estuarine regions of the tropical and subtropical coastlines (Duke 1992). The mangrove is also defined as assemblages of salt tolerant trees and shrubs. Mangroves are taxonomically heterogenous and more related to environment than terrestrial or aquatic plants.

The term 'mangroves' is used to define both a group of plants and a community. To avoid confusion, Macnae (1968) proposed the term 'mangal' for the mangrove forest community, and the term 'mangroves' for the plants. Other terms for the mangrove forest community are coastal woodland, oceanic rainforest, intertidal forest, tidal forest, blue carbon forest, mangrove forest, and mangrove swamp.

Generally, mangrove species are grouped into two types: Exclusive and Non-Exclusive. The exclusive species are limited to the mangrove habitat, and they are also referred to as strict mangroves, obligate mangroves or true mangroves. The non-exclusive species are mainly distributed in a terrestrial or aquatic habitat, but they also occur in the mangrove habitats and they are referred to as semi-mangroves, back mangroves or mangrove associates. Tansley and Fristch (1905) are the first to introduce the criteria to classify mangrove species into true mangroves and mangrove associates. Tomlinson (1986) used rigid criteria to distinguish true mangroves from mangrove associates. In his criteria, true mangroves possess all or most of the following features:

1. They occur exclusively in mangal.
2. They play a major role in the structure of the community and have the ability to form pure stands.
3. They have morphological specializations – especially aerial roots and specialized mechanisms of gas exchange.
4. They have physiological mechanisms for salt exclusion and/or excretion.
5. They have viviparous reproduction.
6. They are taxonomically isolated from terrestrial relatives. The strict mangroves are separated from their nearest relatives at least at the generic level, and often at the sub-family or family level.

Peculiarities of Mangroves

The mangrove flora is considerably distinct from other plants in its peculiar morphological, anatomical, and physiological features. Located in the transition zone between land and sea, the mangroves adapt to a wide range of environmental conditions of salinity, tidal inundation, anoxic sediment, soil erosion, accretion, tidal currents and cyclonic storms. Key adaptations of mangrove are as follows

1. Salt exclusion through roots
2. Salt excretion in leaves
3. Salt accumulation or succulence in leaves
4. Xeromorphic leaf adaptations
5. Specialized aerenchyma root
6. Water dispersed propagules

Salt exclusion and salt excretion are the important mechanisms of salt regulation in mangroves. *Rhizophora* and other members of family Rhizophoraceae have a well-developed mechanism of ultra-filtration in their roots enabling only selective absorption of ions while extracting water from the soil. They may retain a low internal salinity by means of salt-excluding mechanisms in the roots. *Avicennia* sp., *Aegiceras* sp., and several other species have salt glands on their leaves which secret salt. Salt accumulation is also another salt-regulatory mechanism found in species of *Lumnitzera* and *Excoecaria* which accumulate salts in leaf vacuoles and become succulent.

Water dispersed propagules of *Rhizophora mucronata*

Salt exclusive stilt root of *Rhizophora mucronata*

Salt excreting leaves of *Avicennia officinalis*

Specialized aerenchyma roots of *Sonneratia alba*

Stilt roots of *Rhizophora apiculata*

Succulent salt accumulating leaves of *Lumnitzera littorea*

Structural diversity in Mangroves

Nypa fruticans

Growth form varies with mangrove species: trees (*Sonneratia caseolaris, Bruguiera gymnorhiza*), shrubs (*Aegiceras corniculatum*), trunk-less palm (*Nypa fruticans*), and ground fern (*Acrostichum aureum, Acrostichum speciosum*). Further the plants vary as columnar and erect (*Bruguiera parviflora*), spreading and sprawling (*Acanthus* spp., *Scyphiphora hydrophylacea*), and multi-stemmed (*Ceriops tagal*). Growth form also varies within the same species (*Lumnitzera littorea* and *Rhizophora* spp.) having both erect and tangled thicket form. In general, the plants on the edges of mangrove stands have lower limbs and foliage, and their stems are typically sprawling and sinuous, rather than erect and straight. Some mangrove species form closed canopies with combination of different species (*Avicennia marina, Bruguiera gymnorhiza, Bruguiera parviflora, Rhizophora apiculata* and *Xylocarpus* spp.), while others are commonly found as under-canopy plants beneath the closed canopy (*Aegiceras corniculatum, Cynometra iripa, Acanthus* spp., *Acrostichum* spp.).

Rhizophora mucronata

Tangled thicket form of *Rhizophora apiculata*

Tall strands of *Rhizophora spp*

***Acrostichum aureum* (mangrove fern)**

Sprawling shrub of *Scyphiphora hydrophylacea*

Spreading shrub of *Acanthus ilicifolius*

Leaf diversity in Mangroves

Glossy leaves of
Scyphiphora hydrophylacea

Mangrove leaves are leathery and glossy in appearance due to the presence of thick cuticle with obscure leaf veins. Generally, leaves are moderate sized with xerophytic adaptation to cope up with dry and saline conditions. Mangrove leaves exhibit a wide range of variations: simple (*Avicennia* spp., *Rhizophora* spp., *Sonneratia* spp.), compound (*Cynometra* spp., *Xylocarpus* spp.) or pinnate (*Acrostichum* spp.). Leaf shape varies from apiculate (*Rhizophora* spp.), and ovate (*Sonneratia alba*) to spathulate (*Lumnitzera* spp.). Leaf tip ranges from pointed (*Avicennia marina*), round (*Avicennia officinalis*), emarginate (*Lumnitzera* spp.) to mucronate tip (*Rhizophora* spp. and *Sonneratia* spp.). Leaf margin is entire (*Rhizophora* spp.), serrate (*Excoecaria agallocha*), or spiny (*Acanthus ilicifolius*). Leaf surface is smooth and glabrous (*Rhizophora* spp., *Bruguiera* spp., *Ceriops* spp.) or pubescent (*Heritiera* spp.), or with salt excreting glands (*Aegiceras corniculatum*), or milky sap (*Excoecaria* spp.). Leaf size is large (*Heritiera* spp.) or small (*Pemphis acidula*).

Compound leaves of *Xylocarpus moluccensis*

Leaves of *Avicennia officinalis* with rounded tip

Pinnate leaves of *Acrostichum aureum*

Ovate leaves of *Sonneartia alba*

Mucronate tip of *Rhizophora mucronata*

Mucronate tip of *Sonneratia alba*

Leaves of *Excoecaria agallocha* with serrated leaf margin

Small pubescent simple leaves of *Pemphis acidula*

Bark diversity in Mangroves

Bark characteristics vary with mangrove species. Bark texture is smooth (*Avicennia marina*), flaky (*Xylocarpus moluccensis*), fissured (*Lumnitzera* spp.), pustular (*Excoecaria agallocha*), friable and crumbly (*Bruguiera gymnorhiza*), or crocodile skin (*Rhizophora apiculata*). Bark colour is red (*Rhizophora stylosa*), dirty white (*Avicennia marina*), brown (*Bruguiera parviflora*), and black (*Lumnitzera* spp.).

Rhizophora apiculata

Xylocarpus moluccensis

Lumnitzera littorea

Heritiera littoralis

Avicennia marina

Sonneratia ovata

Xylocarpus granatum

Excoecaria agallocha

Bruguiera parviflora

Bruguiera gymnorhiza

Rhizophora stylosa

Buttress and Planks root of *Xylocarpus granatum*

Root diversity in Mangroves

Above-ground roots are the most striking structures of mangroves. They help for anchorage in loose muddy soil, absorbing nutrients and exchanging gases in oxygen-deficit soil condition. The above-ground roots are the pneumatophores with pencil-like (*Avicennia* spp.), stiff conical (*Xylocarpus moluccensis*), flexible conical (*Sonneratia alba*), or elongated conical (*Sonneratia caseolaris*) structures; the knee roots are thick and knobbly (*Bruguiera* spp.), or thin and wiry (*Lumnitzera littorea*); the buttresses as sinuous planks (*Xylocarpus granatum*, *Heritiera littoralis*) and erect fins (*Bruguiera gymnorhiza*); and the stilt and prop roots (*Rhizophora* spp.). Stilt roots of *Rhizophora* species play a significant role in its special ultra-filtration system that can filter nearly 90% of sodium ions in saline water. The root possesses a hierarchical, triple layered pore structure in the epidermis, and most Na+ ions are filtered at the first sub-layer of the outermost layer. The high blockage of Na+ ions is attributed to the high surface zeta potential of the first layer (−91.4 ± 0.93 mV; Kim et al. 2016a, b). The second layer, which is composed of macroporous structures, also facilitates Na+ ion filtration.

Pneumatophores of *Avicennia marina*

Pneumatophores of *Sonneratia alba*

Prop root of *Rhizophora stylosa*

Stilt root of *Rhizophora mucronata*

Planks root of *Heritiera littoralis*

Pneumatophores of *Xylocarpus moluccensis*

Knee root of *Bruguiera gymnorhiza*

Flower diversity in Mangroves

Flowers are the most attractive part of the plant. Like other flowering plants, the mangrove plants have flowers to attract pollinators that facilitate the transfer of pollen grains from male to receptive female reproductive part. Mangrove flowers significantly vary with species. There are combined sex flowers (*Rhizophora* spp., *Sonneratia* spp., *Avicennia* spp.), separate sex flowers (*Heritiera littoralis*), or separate sex trees (*Excoecaria agallocha*). The flower colour is white (*Sonneratia alba, Lumnitzera racemosa*), red (*Sonneratia caseolaris, Lumnitzera littorea*), purple (*Acanthus ilicifolius*), or orange yellow (*Avicennia marina*). The flower is large (*Dolichandrone spathacea*), medium (*Bruguiera gymnorhiza*) and, or small (*Bruguiera parviflora, Bruguiera cylindrica*).

Sonneratia caseolaris

Heritiera littoralis

Male flower of *Excoecaria agallocha*

Aegiceras corniculatum

Young male flowers *of E. agallocha*

Cynometra iripa

Pemphis acidula

Sonneratia alba

Xylocarpus granatum

Reproductive diversity in Mangroves

Twin propagules of *Rhizophora mucronata*

Vivipary in *Rhizophora mucronata*

Mangroves are pollinated almost exclusively by animals (bees, small insects, moths, bats, and birds), while *Rhizophora* spp., are primarily self-pollinated. Vivipary is the most important adaptation of many mangrove species. It is defined as a condition in which embryo germinates while still attached to the plants. The embryo has no dormant stage but grows out of seed coat and fruit before detaching from the plant. However not all mangroves have this vivipary. Several species do not have any viviparity, while many species have lesser degree of Vivipary (cryptovivipary). Type of propagules varies with mangrove species: seeds (*Xylocarpus* spp., *Sonneratia* spp.), cryptovivipary (*Avicennia* spp.) and vivipary (*Rhizophora* spp., *Bruguiera* spp., *Ceriops* spp., *Kandelia* spp.). Structures of propagule are drupe (*Sonneratia* spp.), capsules (*Excoecaria agallocha*), hypocotyls (*Rhizophora* spp., *Bruguiera* spp., *Ceriops* spp., *Kandelia candel*), cotyledons (*Avicennia* spp., *Lumnitzera* spp.), fleshy (*Sonneratia ovata*, *Sonneratia caseolaris*), woody (*Xylocarpus* spp.), persistent style (*Sonneratia* spp., *Avicennia* spp.), keeled pods (*Heritiera littoralis*) or elongated pointed radicle (*Rhizophora mucronata*, *Rhizophora stylosa*).

Three lobed Fruits of *Excoecaria agallocha*

Large seeds of *Xylocarpus granatum*

Cryptovivipary of *Avicennia officinalis*

Sickle shaped seeds of *Sonneratia alba*

Pod like fruits of *Heritiera littoralis*

Part 2: Mangrove floristics of the World

Origin and diversification of Mangroves

Mangroves are quite old, possibly arising just after the first angiosperms (Duke 1992). Based on the fossil evidence and recent molecular studies, it is believed that mangroves originated along the shorelines of ancient Tethys Sea and there was a dispersal along the Tethys seaway westward into the Atlantic and further into the east Pacific (long before the closure of the Panama Isthmus) and eastward into South east Asia and Australia during the Late Cretaceous or early Eocene around 80 million years ago (Duke 2006). *Nypa* seems to be earliest fossil evidence. However, prior to closure of Tethys sea, around 40–55 million years ago, the mangrove genera - *Nypa*, *Acrostichum*, *Rhizophora*, *Avicennia*, *Pelliciera*, *Sonneratia*, *Bruguiera*, *Ceriops*, *Heritiera*, and *Aegiceras* - were present around shoreline (Duke 2017). The closure of Tethys Seaway by the mid-Tertiary (34–50 Million years ago) by northward movement of African landmass, created a physical barrier that completely terminated the Atlantic East Pacific (AEP) and Indo-West Pacific (IWP) exchange route, and resulted in subsequent diversifications within the AEP and IWP regions influenced by region-specific events (Ellison et al. 1999; Duke 2006). The plate tectonics and other vicariant events like Pleistocene sea level rise are attributed to the current distribution pattern and disjunct distributions of species within genera. Duke (2017) hypothesised two separate westward dispersal routes in AEP: (i) a southern path along the coasts of West Africa, South America and, (ii) a northern path along the North American east coast, whereas he hypothesised two isolated pathways in IWP towards eastward dispersal: (i) a northern path along the coast to the Middle East, Southeast Asia and, (ii) a southern path along the coasts of East Africa, India, and Australia. Further he noted that collision of India and Australia with Middle East and Southeast Asia respectively, resulted complication in dispersal route. Based on the available records it is presumed that diversification of mangrove plants has largely been driven by three key processes of speciation viz., Allopatric speciation, Peripatric speciation and Parapatric speciation (Duke 2017).

Floristic Composition

Mangrove floristic composition has been extensively reviewed in the past (e.g. Saenger et al. 1983; Tomlinson 1986; Ricklefs and Latham 1993; Duke 1992; Field 1995; Duke et al. 1998; Ellison et al. 1999; Kathiresan and Bingham, 2001; Saenger 2002; Wang et al.2003; Spalding et al. 2010; Polidoro et al. 2010). The information of species composition and distribution is a prerequisite for effective conservation and management. Globally there are 85 true mangrove species (including 14 natural hybrids) belonging to 30 genera under 17 botanical families (Table 1). However, uncertainty prevails due to lack of clear-cut demarcation between true mangroves and their associate species. For instance, Duke (2017) excluded the species namely *Acanthus volubilis*, *Acanthus xiamenensis*, *Phoenix paludosa*, *Brownlowia argentata*, *Heritiera globosa*, *Aglaia cucullata*, and *Excoecaria indica* from the list of global mangroves. All these species except *Heritiera globosa* are retained in our text as they constitute important vegetation in mangroves of Indo-Malesia. *Heritiera globosa*, once considered as mangrove species (Duke 1992; Giesen et al. 2006) is excluded in our text as it is found only at the upper reaches of freshwater rivers. Further, certain species namely *Muellera moniliformis*, *Barringtonia racemosa*, *Crenea patentinervis*, *Pavonia paludicola*, and *Pavonia rhizophora*, counted as mangroves by Duke (2017) are not considered here. Similarly, *Rhizophora × beristyla* is also not included in our text as it was considered as synonym of *R. mangle* (Hou 1960; Breteler 1977), and later it was lectotypified as synonym of *R. × harrisonii*. Out of the 64 orders of angiosperms, as per Angiosperm Phylogeny IV (APG IV), 10 orders (Arecales, Caryophyllales, Malpighiales, Fabales, Myrtales, Malvales, Sapindales, Laminales, Gentianales and Ericales) represents global mangrove species (Fig. 1).

Table 1. Mangrove species of the world ([1]Present in Atlantic-East Pacific mangroves; [2]Present in both Atlantic-East Pacific and Indo-West Pacific mangroves)

FAMILY	GENUS	SPECIES
ACANTHACEAE	*Acanthus*	*Acanthus ebracteatus*
		Acanthus ilicifolius
		Acanthus volubilis
		Acanthus xiamenensis
	Avicennia	*Avicennia alba*
		Avicennia bicolor [1]
		Avicennia germinans [1]
		Avicennia schaueriana [1]
		Avicennia integra
		Avicennia marina
		Avicennia officinalis
		Avicennia rumphiana
ARECACEAE	*Nypa*	*Nypa fruticans*
	Phoenix	*Phoenix paludosa*
BIGNONIACEAE	*Dolichandrone*	*Dolichandrone spathacea*
	Tabebuia	*Tabebuia palustris* [1]
COMBRETACEAE	*Conocarpus*	*Conocarpus erectus* [1]
	Laguncularia	*Laguncularia racemosa* [1]
	Lumnitzera	*Lumnitzera littorea*
		Lumnitzera racemosa
		Lumnitzera × rosea
EBENACEAE	*Diospyros*	*Diospyros littorea*
EUPHORBIACEAE	*Excoecaria*	*Excoecaria agallocha*
		Excoecaria indica
		Excoecaria ovalis
FABACEAE	*Cynometra*	*Cynometra iripa*
	Mora	*Mora oleifera* [1]
LYTHRACEAE	*Pemphis*	*Phemphis acidula*
	Sonneratia	*Sonneratia alba*
		Sonneratia apetala
		Sonneratia caseolaris
		Sonneratia griffthii
		Sonneratia lanceolata
		Sonneratia ovata
		Sonneratia × gulngai
		Sonneratia × urama
		Sonneratia × hainanensis
		Sonneratia × zhongcairongii
MALVACEAE	*Brownlowia*	*Brownlowia argentea*
		Brownlowia tersa
	Camptostemon	*Camptostemon philippinensis*
		Camptostemon schultzii
	Heritiera	*Heritiera fomes*
		Heritiera littoralis
MELIACEAE	*Aglaia*	*Aglaia cucullata*
	Xylocarpus	*Xylocarpus granatum*
		Xylocarpus moluccensis
PRIMULACEAE	*Aegiceras*	*Aegiceras corniculatum*
		Aegiceras floridum
MYRTACEAE	*Osbornia*	*Osbornia octodonta*
PLUMBAGINACEAE	*Aegialitis*	*Aegialitis annulata*
		Aegialitis rotundifolia
PTERIDACEAE	*Acrostichum*	*Acrostichum aureum* [2]
		Acrostichum danaeifolium [1]
		Acrostichum speciosum
RHIZOPHORACEAE	*Bruguiera*	*Bruguiera cylindrica*
		Bruguiera exaristata
		Bruguiera × hainesii
		Bruguiera gymnorhiza
		Bruguiera parviflora
		Bruguiera sexangula
		Bruguiera × rhynchopetala
		Bruguiera × dungarra
	Ceriops	*Ceriops australis*
		Ceriops decandra
		Ceriops pseudodecandra
		Ceriops tagal
		Ceriops zippeliana
	Kandelia	*Kandelia candel*
		Kandelia obovata
	Rhizophora	*Rhizophora apiculata*
		Rhizophora mangle [1]
		Rhizophora samoensis [2]
		Rhizophora mucronata
		Rhizophora racemosa [1]
		Rhizophora stylosa
		Rhizophora × lamarckii
		Rhizophora ×annamalayana
		Rhizophora × mohanii
		Rhizophora ×selala
		Rhizophora × tomlinsonii
		Rhizophora × harrisonii [1]
RUBIACEAE	*Scyphiphora*	*Scyphiphora hydrophylacea*
TETRAMERISTACEAE	*Pelliciera*	*Pelliciera rhizophorae* [1]
		Pelliciera benthami [1]

Figure 1: Phylogenetic position of Mangrove families as per APG IV

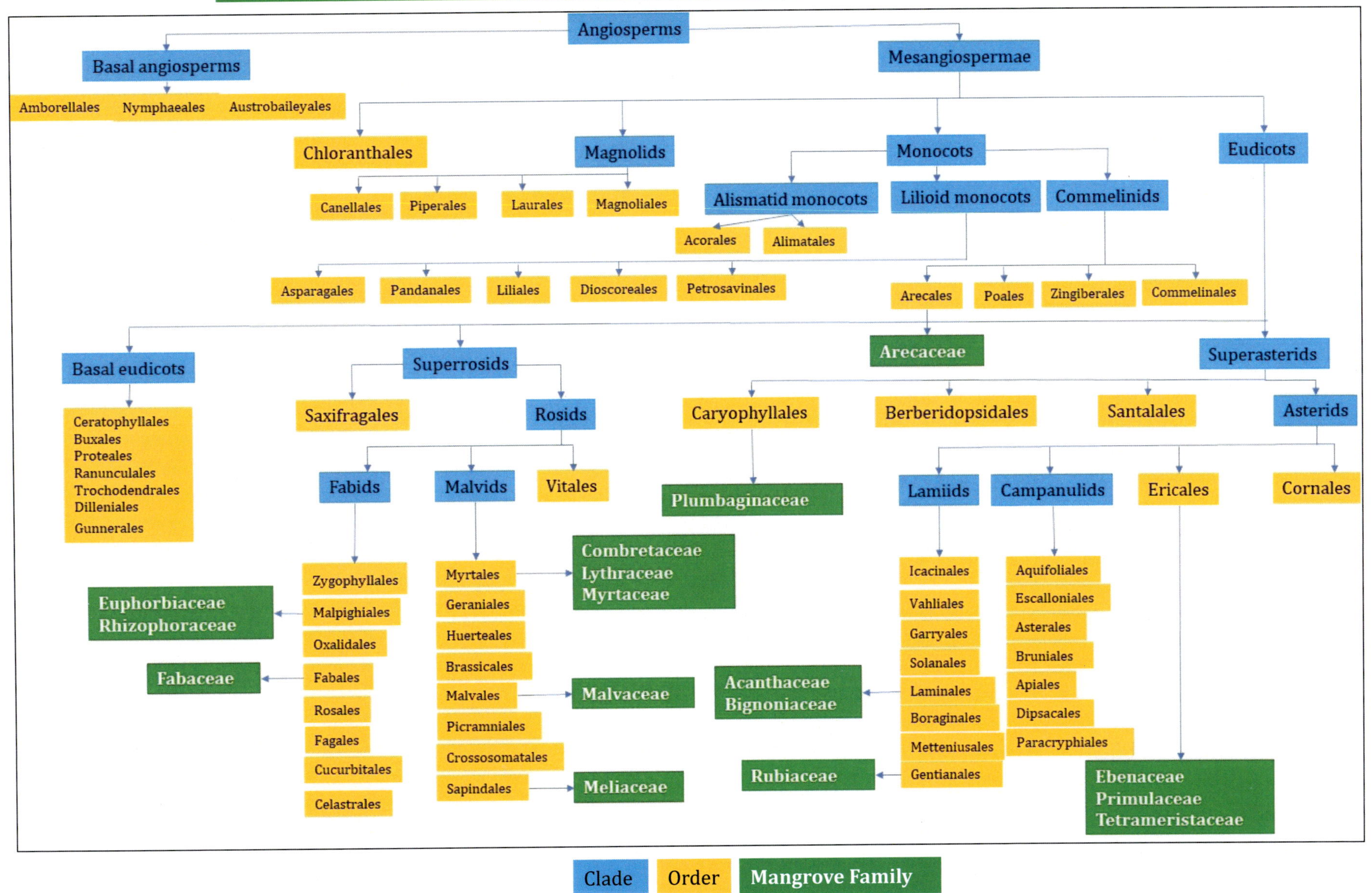

Global Distribution of Mangroves

Mangroves are distributed in sheltered tropical and subtropical coastlines within latitudes of 32°N and 38°S. They occupy an area of 137,760 km^2 spanning in 118 countries and territories (Giri et al. 2011). The mangroves are sensitive to freezing and chilling temperatures, and they can survive only at the mean winter sea surface temperature of at least 20 °C. However, now mangroves are expanding into temperate regions due to the reduced frequency and intensity of extreme freeze events caused by climate change (Saintilan et al. 2014). Longitudinally there are two main centres of global mangroves viz., Atlantic East Pacific (AEP) and Indo-West Pacific (IWP). AEP includes West America, East America and West Africa, while IWP includes East Africa, Indo-Malesia and Australasia (Fig. 2). Despite identical physical conditions, IWP is biologically diverse with 72 mangrove species, as compared to AEP with only 15 mangrove species (Table 1). Both the IWP and AEP are commonly distributed with two species (*Acrostichum aureum*, *Rhizophora samoensis*) and three genera (*Acrostichum, Avicennia* and *Rhizophora*). All the botanical families except Tetrameristaceae (formerly Pelliceriaceae) are present in both AEP and IWP. Difference in species richness and species composition between the IWP and the AEP is attributed to complete isolation caused by the closure of Tethys seaway and regional differences in the rate of origin of new mangrove lineages. Continuous presence of large areas of continental shelf with islands scattered among shallow tropical seas, resulted by more complex geological history of tectonic movements, and caused high rate of origin of new mangrove lineages in IWP compared to the AEP.

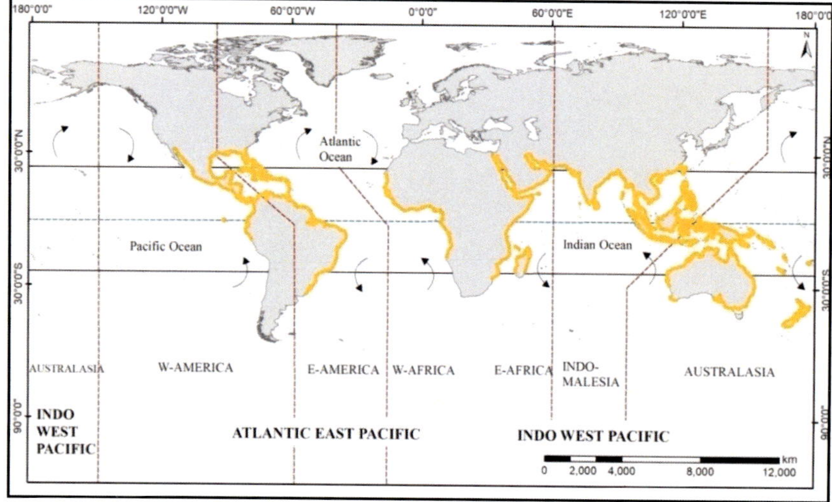

Figure 2: Global distribution of Mangroves. (Reproduced based on Duke 2017)

Distribution of Mangrove species richness in the IWP

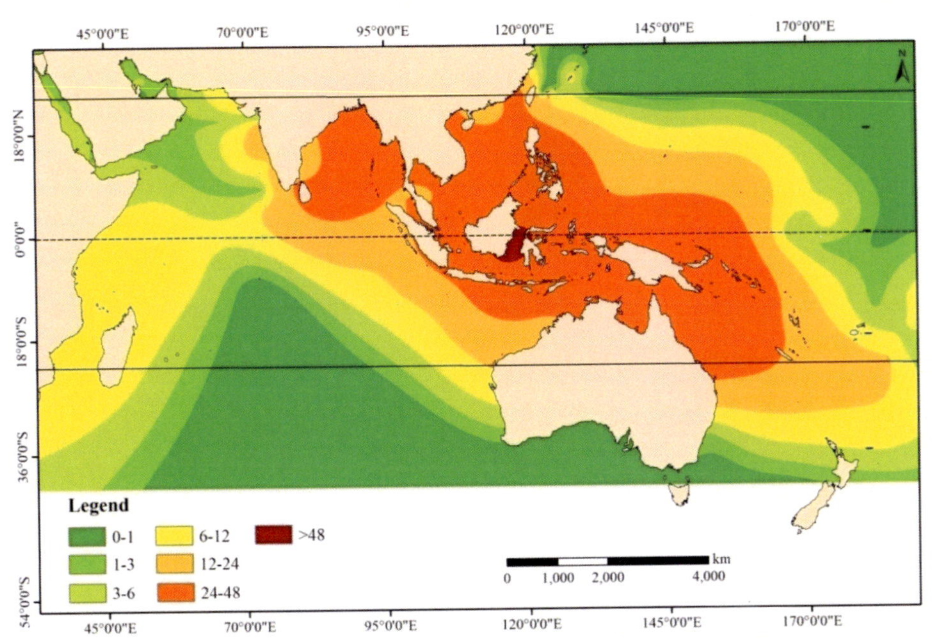

Figure 3: Approximate distribution of mangrove species richness in the Indo-West Pacific. (Based on data from Spalding et al. (2010) in http://www.nies.go.jp/TroCEP/index.html. Species size classes are based on a geometrical progression. Source: Saenger et al. 2019).

Mangrove species richness and mangrove development (e.g. height and extent of mangrove vegetation) decreases with increasing latitude. The latitudinal limits of global mangroves are quite variable; however, they can be broadly related to temperature and/or aridity (Saenger et al. 2019). The distribution of mangrove species richness across the IWP shows maximal richness that occurs along the shorelines of Makassar Strait, between Borneo and Sulawesi in Indonesia. This area has the greatest interdigitation of land and sea in the world, a condition stretching back to the Holocene. While some seaways may have closed due to tectonic or sea-level changes, the area has always maintained some direct connections between the tropical Pacific and tropical Indian oceans as Makassar Strait lies more or less seaward of the Sunda Shelf. High precipitation and relatively low salinities in this area also provided favourable conditions for the development and maintenance of the mangrove niche. Furthermore, cycles in sea level during the numerous late Pliocene and Pleistocene glacial-interglacial intervals have led to appearance or disappearance of islands, the opening or closing of marine corridors and bridges with alternating higher and lower salinities (Voris 2000). In turn, these processes have led to bottleneck and founder effects on the mangrove abundance in the region and led to high rates of speciation. The second identifiable trend in the figure-3 is the attenuation of species numbers with increasing latitude generally related to limiting temperatures. Thus, the rapid loss of species occurs on the east and west coast of Australia, in southern Africa and along the shores of the East China Sea. The rapid decline in species on the northern shores of East Africa and into the Red Sea, Gulf of Oman and Arabian Gulf is certainly due to limiting aridity rather than temperature. An attenuation of species numbers from continental land masses to offshore archipelagos is also discernible although it is less marked.

Genetic Diversity and dispersal barriers of Mangroves

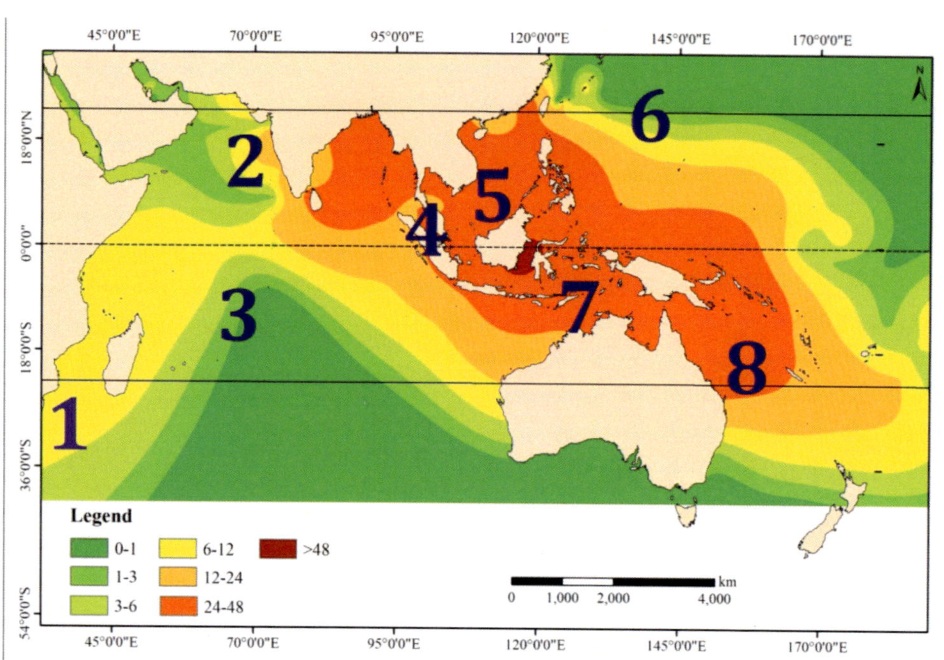

Figure 4: Approximate distribution of mangrove species richness in the Indo-West Pacific with the numbered locations of those areas where restricted gene flow has been demonstrated for several mangrove species 1. South West Indian Ocean; 2. Arabian Sea; 3. Western Central Indian Ocean; 4. Malacca strait; 5. South China Sea; 6. Western Central Pacific Ocean; 7. Timor Sea; 8. Coral Sea (Source: Saenger et al. 2019)

> The MIM mechanism, by relaxing the condition of no gene flow, can promote speciation in many more geographical features than strict allopatry can (He et al. 2018).

Mangroves are once believed to be genetically undifferentiated throughout their range as a result of long-distance dispersal (LDD) of their propagules (Duke et al. 1998; Maguire et al. 2000). However, the potential of LDD is limited due to factors including land barriers and ocean circulations (Dodd and Rafii 2002; Takayama et al. 2013). In the IWP region, the Malay Peninsula plays a major role in promoting the genetic differentiation between populations from the eastern and western coasts (Ng et al. 2015). During the glacial periods, sea level falls caused the Malay Peninsula, Sumatra and Java to get connected and to form the Sunda Land, which halted the exchange of seawater between the Indian Ocean and the South China Sea (SCS) (Wang et al. 1995; Wyrtki 1961). This glacial isolation resulted in genetic divergence between populations from these two oceans, as recently observed in many mangrove species (Li et al. 2016). Other geographical barriers like Torres Strait also significantly influences the genetic divergence (Fig. 4). Thus, most mangrove species exhibit low genetic diversity and strong genetic differentiation between populations (Guo et al. 2018). These studies also revealed the role of vicariance, pleistocene sea level fluctuations, ocean currents, geomorphology, hydrology of estuary in shaping the present distribution and population structure of mangrove species (Dodd et al. 2002; Orsini et al. 2013; Wee et al. 2014; Cerón-Souza et al. 2015; Guo et al. 2018). In general, there is a strong genetic differentiation between the populations of Indo-Malesia and Australasia and within the populations of Indo-Malesia between South China Sea and Indian Ocean. Based on the dynamic role of Strait of Malacca either as a geographical barrier or a conduit of gene flow for mangrove species, He et al. (2018) discovered a novel mechanism of speciation, called "mixing-isolation-mixing" (MIM) cycles.

Mangrove Forest structure

Mangrove forest structure is simple, as compared to other tropical forests. The mangroves are generally with zonation and without understory. The absence of understory is due to combined effects of salinity, flooding, and low light intensity in mangrove habitats (Janzen 1985; Lugo 1986), whereas the zonation pattern is influenced by various biotic and abiotic factors (Bernini and Rezende 2011). The mangrove forest structure is measured by canopy height, basal area, tree density, age/size class distribution and species richness (Oliver and Larson 1996). These measures help to evaluate the development or maturity of a forest ecosystem (Pool et al. 1977; Kangas 2006). The complexity index (*Ic*) and importance value index (Iv) are structural indices used to express the existence of stress in the forest stand and the importance of a tree species within a stand of mixed species, respectively. In general, arid mangroves have low complex index with high stem density, low species richness, low canopy height, and low basal area; whereas, the wet humid mangroves have high complex index with high canopy height, high basal area, and low stem density (Komiyama et al. 2008). All the mangrove habitats of India except Andaman and Nicobar Islands (A & N Islands) have low *Ic* values. *Avicennia* spp. constitute an important mangrove tree species in coastal India; but *Rhizophora* spp., are the important tree species in A & N Islands (Ragavan et al. 2019). Local environmental conditions such as nutrient availability, salinity and hydrology along with regional climate and geomorphology are determinants of mangrove ecotypes, from scrub (< 3 m) to tall (> 15 m) forest stands (Twilley et al. 2017). Recent remote sensing on global analysis of mangrove canopy heights revealed that precipitation, temperature and cyclone explain 74% of the global trends in maximum canopy height, with other geophysical factors that influence the observed variability at local and regional scales. The tallest mangrove forest with 62.8 m is found in Gabon, equatorial Africa (Simard et al. 2019). The dwarf nature of mangrove stands is attributed to limited supply of N, P and Fe from soil (Reef et al. 2010).

Zonation in Mangroves

Mangrove zonation—distribution of tree species in monospecific bands parallel to the shoreline—is a classical feature of mangrove forests, presumed to be present in all the mangroves worldwide (Chapman 1975; Snedaker 1982). This mangrove zonation is due to plant succession upon land building, geomorphological factors, physiological adaptation, dispersal pattern of propagules, predation pressure on propagules, and interspecific competitions across the intertidal zone (Smith 1992). However, not all researchers have supported this classical view of mangrove zonation (Thom 1967; Thom et al. 1975; West 1956; Macnae and Kalk 1962). The mangrove zonation is also observed to be absent in various regions of the globe as evident by lack of strong relationships between species distribution and edaphic gradient (Bunt et al. 1991; Bunt 1996, 1999; Bunt and Bunt 1999; Bunt and Stieglitz 1999; Ellison et al. 2001; Schmiegelow and Gianesella 2014). Thus, a classical zonation pattern in mangrove forest tends to be an exception rather than the rule (Smith 1992).

In Indian mangroves, three zones—proximal, middle, and distal zones, followed by a fourth littoral zone—are described on the basis of distribution pattern of mangrove species (Sidhu 1963; Blasco 1977; Dagar 1982; Dagar et al. 1991; Dagar and Singh 1999). This is only based on qualitative assessments with biased assumption of zonation. Further, the reports on mangrove species zonation do not accurately reflect the complex vegetation patterns of the forest (Ellison et al. 2001). In general, different species of mangroves respond differently to underlying edaphic gradients, and many mangrove species can grow in a broad range of conditions found across the intertidal zone. Hence, understanding the site-specific habitat suitability of mangrove species and interconnection between various components of mangrove ecosystem is necessary to underpin the true science behind mangrove zonation.

Part 3: Mangroves of India

Distribution of Mangroves in India

Mangrove forests in India are found along the coastline of 9 States and 4 Union Territories (Fig. 5). The mangrove habitats are broadly classified into three namely, Deltaic (Eastern Coast Mangroves), Estuarine & Backwater (Western Coast Mangroves) and Insular mangroves (A & N Islands) (Mandal and Naskar 2008). India has a total mangrove cover of 4975 km²; of which, 57% is present along the east coast (Bay of Bengal); 31% along the west coast (Arabian Sea) and the remaining 12% in the Andaman and Nicobar Islands (FSI 2019).

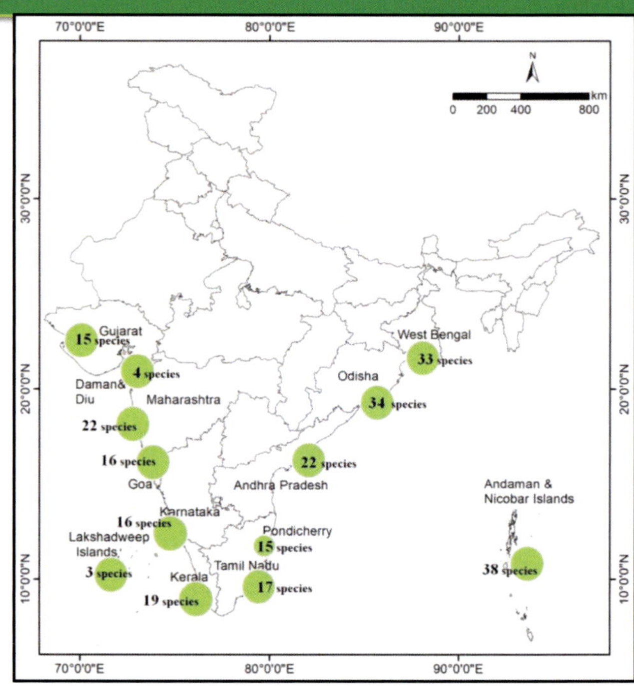

Figure 5: Distribution of mangroves in India

The mangrove cover is larger and more widespread on the east coast than that on the west coast. This difference in mangrove cover can be attributed to distinctive geo-morphological settings, 1) the east coast has large estuaries with deltas formed due to runoff and deposition of sediments, whereas the west coast has funnel-shaped estuaries with absence of deltas; and 2) the east coast has gentle slopes with extensive flats for mangrove colonization, whereas the west coast has steep slopes (Kathiresan 2010). The mangroves of A & N Islands are probably the best developed in India in terms of their density and growth (Ragavan et al. 2019). Their irregular and deeply indented coastline results in innumerable creeks, bays and estuaries which facilitate the development of extensive and luxuriant growth of mangrove forests with a high degree of biodiversity.

Mangroves floristics of India

Indian mangroves represent 46 true mangrove species, including 4 natural hybrids, belonging to 14 families and 22 genera. The East coast has 40 mangrove species under 14 families and 22 genera. The West coast has 27 species belonging to 11 families and 16 genera, whereas the A & N Islands have 38 species under 13 families and 19 genera. Among the 13 States/Union Territories, mangrove diversity is highest in the A & N Islands. Species namely *Rhizophora × lamarckii, Lumnitzera littorea*, *Sonneratia ovata*, *S. lanceolata*, *S. × urama* and *S. × gulngai* are restricted to A & N Islands. In mangrove diversity, India is the third richest country in the world (after Indonesia and Australia; Ragavan et al. 2016).

True mangrove species of India

Acanthus ebracteatus Vahl
Acanthus ilicifolius L.
Acanthus volubilis Wall.
Acrostichum aureum L.
Acrostichum speciosum Willd.
Aegialitis rotundifolia Roxb.
Aegiceras corniculatum (L.) Blanco
Aglaia cucullata (Roxb.) Pellegr.
Avicennia alba Blume
Avicennia marina (Forssk.) Vierh
Avicennia officinalis L.
Brownlowia tersa (L.) Kosterm.
Bruguiera cylindrica (L.) Blume
Bruguiera gymnorhiza (L.) Lam.
Bruguiera parviflora Wight & Arn. Ex Griff
Bruguiera sexangula (Lour.) Poir.
Ceriops decandra (Griff.) Ding Hou
Ceriops tagal (Perr.) C. B. Rob.
Cynometra iripa Kostel.
Dolichandrone spathacea (L.f.) Baill. ex
Excoecaria agallocha L.
Excoecaria indica (Willd.) Muell.-Arg.
Heritiera fomes Buch.-Ham
Heritiera littoralis Dryand.
Kandelia candel (L.) Druce
Lumnitzera littorea (Jack.) Voigt
Lumnitzera racemosa Willd.
Nypa fruticans (Thunb.) Wurmb
Pemphis acidula J.R. Forst.
Phoenix paludosa Roxb.
Rhizophora × annamalayana Kathiresan
Rhizophora × lamarckii Montrouz
Rhizophora apiculata Blume
Rhizophora mucronata Lam.
Rhizophora stylosa Griff.
Scyphiphora hydrophylacea C.F.Gaertn
Sonneratia × gulngai N.C. Duke
Sonneratia × urama N.C. Duke.
Sonneratia alba Sm.
Sonneratia apetala Buch.-Ham.
Sonneratia caseolaris (L.) Engl.
Sonneratia griffithii Kurz
Sonneratia lanceolata Blume
Sonneratia ovata Backer
Xylocarpus granatum J. Konig
Xylocarpus moluccensis (Lam.) M. Roem.

Description of Mangrove Species of India

Genus: *Acanthus*

The genus *Acanthus* is distinguished from related genera by its spiny leaves, spicate inflorescences, presence of bracteoles and uniform anther. The genus *Acanthus* is represented by four species in mangroves. Of these, *A. xiamenensis* is endemic to China and all the other species are common in IWP region. However, the taxonomical identity of *A. xiamenensis* in China is not clear; for instance, Wang and Wang (2007) have treated *A. xiamenensis* and *A. ilicifolius* as the same species. In India three species viz., *Acanthus ilicifolius*, *A. ebracteatus*, and *A. volubilis* are known. *Acanthus* species are often found in landward edges of mangroves just above the high tide mark, and also occur in inner mangroves as understory. Individuals under shade are often not serrated.

Key to *Acanthus* species

1. Bracteoles present 2
   Bracteoles absent or minute...........................3
2. Inflorescences terminal and axial, flowers light blue to dark or violet (rarely white), stem axial spines present or absent, if present always facing upward............***A. ilicifolius***

3a. Inflorescences terminal, flowers white, stem axial spine absent................***A. volubilis***

3b. Inflorescences terminal, flowers white, stem axial spines facing downwards ....***A. ebracteatus***

Acanthus ilicifolius
Spineless leaves with absences of stem axial spines

Acanthus ilicifolius
Spineless leaves with presence of stem axial spines

Acanthus ilicifolius L., Sp. Pl. 2: 639 (1753).

Acanthus ilicifolius is most common in Indian mangroves. Often grows in sheltered mangrove areas, landward margins and back waters. There is considerable variation in leaf form and leaf serration. *Acanthus ilicifolius* is distinguished from *A. ebracteatus* by its purple colour flowers, presence of both axial and terminal inflorescences, presence of bracteoles and upward facing stem axial spines at the base of the petioles.

Terminal inflorescences

KEY CHARACTERS

- Leaves are simple opposite, lanceolate, narrowed at base, serrate margins armed with spines, population under shade are less serrated.
- Stem smooth, stem axial spines always facing upward
- Spicate inflorescences either axial or terminal.
- Flowers are purple in color, rarely white
- Bract single and bracteoles present.
- Fruits ovoid oblong (capsule like)

Species feature

Axial spicate inflorescences

Stem axial spines facing upward

Prominent bracteoles and bract

Purple color flower with hairy stamens

Acanthus ilicifolius L.

Taxonomy

Shrub: height up to 3m (A). *Stem*: thick, green, light green or purple, sparsely branched and stem axial spine either present or absent, if present always facing upward (E). *Roots*: occasionally above ground or prop roots on lower parts of reclining stem (D). *Leaves*: simple, opposite, lanceolate to broadly lanceolate, margin either entire or spiny and dentate, leaf base attenuate, leaf tip acute and narrowly pointed with or without spiny edge, presence of spines with greater sunlight and exposure, size variable, 6–30 × 1.5–6 cm, ratio of length to width is greater than 2; petiole short, green, 0.5–2 cm long. *Inflorescences*: both terminal (B) and axial (C), terminal inflorescences longer than axial, up to 15 cm long, axial inflorescence smaller than terminal, 5–10 cm long. *Mature Flower bud*: ellipsoidal, 3–3.5 cm long; bract single, 0.8–1 × 0.5–0.7cm (H); bracteoles 2, lateral, 0.5–0.8 × 0.2–0.4 cm (H); calyx four lobed, outer two larger in size, 1.3–1.5 × 0.8–1.2 cm, enclosing the flower bud, inner lateral lobes narrow, 1 × 0.5–0.8 cm, enclosed by upper and lower lobes; corolla purple or deep purple, rarely white with dark blue median band (B & I), 3–4 ×2–2.5 cm; stamens 4, 2–2.5 cm long, sub equal with thick hairy connectives; anther medifixed each with 2 cells aggregated around the style, 0.8–1cm long, densely ciliated (F); ovary bi-locular with 2 superimposed ovules in each loculus; style enclosed by stamens, 2.5–3 cm long, capitate to pointed stigma exposed (G & I). *Mature fruits*: 4 seeded capsules, ovoid, green, shiny, smooth, 3–3.5 × 1 cm (J).

Morphological features of *Acanthus ilicifolius. (A)* Habit; (B) inflorescences; (C) buds with bract and bracteoles; (D) stem base with stilt root; (E) upward facing stem axial spines; (F) anthers; (G) style; (H) mature bud; (I) purple flower; (J) fruit.

Distribution

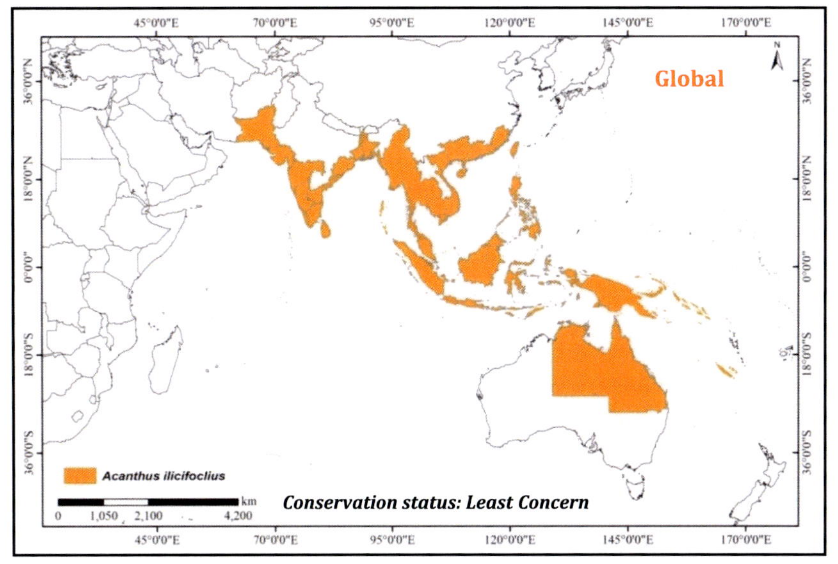

Habitat and Ecology

Often found in landward edges of mangroves just above the high tide mark, and also occur in inner mangroves as understory. Individuals under shade are not serrated.

Phenology

Flowering February to March; Fruiting April to May.

Notes

Despite wide variation in leaf shape, serration and stem axial spine, the presence of prominent bract and bracteoles is most consistent in *A. ilicifolius*. In the absence of flowering and fruiting, the spineless form of *A. ilicifolius* is often misidentified as *A. volubilis* (e.g. Thothathri 1962; Dam Roy et al. 2009). Flower colour of *A. ilicifolius* is sometimes white, instead of violet (Jayatissa et al. 2002), so this species is also misidentified as *A. ebracteatus*. For instance, *Acanthus albus* reported from Sundarbans by Debnath et al. (2013) and *A. ebracteatus* from Bhitarkanika by Pradan et al. (2016) are in fact white flowered variant of *A. ilicifolius*.

Acanthus ebracteatus Vahl, Symb. Bot. (Vahl) 2: 75, t. 40 (1791). Acanthaceae

Acanthus ebracteatus is rare in India and it has restricted distribution in A & N Islands, Kerala, Puducherry, Odisha and Sundarbans. However, identity and locality of this taxon in mainland India remains elusive. This species is often confused with *Acanthus ilicifolius*. It is distinguished from *A. ilicifolius* by its white colour flowers, absence of bracteoles, terminal inflorescences and downward facing stem axial spines. Found in landward edges of mangroves just above the high tide mark, and also in inner mangroves as understorey. Individuals under shade are less serrated.

KEY CHARACTERS

- Shrub growing in brackish water
- Leaves dark green with serrate margins armed with spines
- Stem smooth, stem axial spines facing downward
- Inflorescences terminal, spicate
- Bract single minute and bracteoles absent or obscure
- Flowers white in colour, flower size smaller than *A. ilicifolius*
- Fruits green and oblong

Species feature

Obscure bracteoles

Spicate terminal inflorescences with white flowers

Capsule like fruits without bracteoles

Stem axial spine facing downwards

Stamen with hairy anther

Acanthus ebracteatus vahl

Taxonomy

Shrub: height to 2 m (A). *Stem*: thick, grey colour, sparsely branched, stem axial spines always present and facing downward (C). *Roots*: occasionally above ground or prop roots on lower parts of reclining stem (B). *Leaves*: simple, opposite, broadly elliptic to lanceolate, 10–20 × 3–6 cm, leaf tip acute to obtuse with or without spiny edge, base attenuate, margin either or spiny and dentate, presence of spines with greater sunlight and exposure; petiole short 0.5–1.5 cm long. *Inflorescence*: always terminal (D). *Mature Flower bud*: ellipsoidal, 1.8–2.2 cm long; bract single not persistent (E, K), 0.3–0.5 cm long, bracteoles absent or obscure, 0.3–0.5 cm long, apex acute (F); calyx four lobed (L), 0.6–0.8 cm long, upper lobe larger than lower lobe (I), enclosing the flower bud, lateral lobes narrow, enclosed by upper and lower lobes; corolla white, with purple strip in middle of lower lip (H), 0.9–1.5 cm long; stamens 4, 1.2–1.5 cm long, sub equal with thick hairy connectives; anther medifixed each with 2 cells aggregated around the style, 0.3–0.5 cm long (M); ovary bilocular with 2 superimposed ovules in each loculus; style enclosed by stamens, 0.8–1.2 cm long, capitate to pointed stigma exposed (G). *Mature fruit*: 4 seeded capsule, ovoid, green, shiny, smooth, 2–2.5 ×1 cm (J).

Distribution

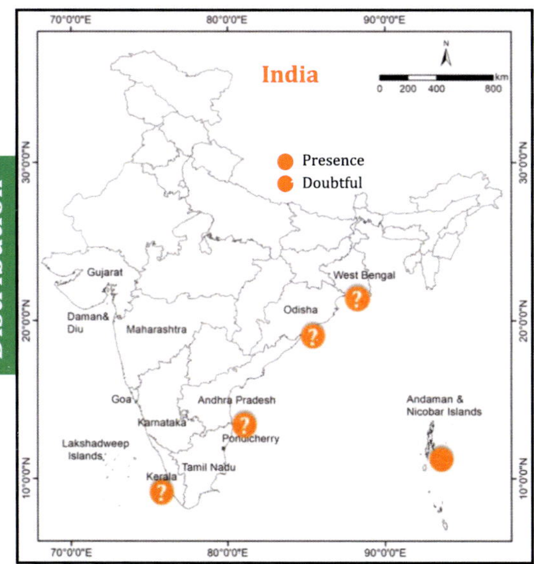

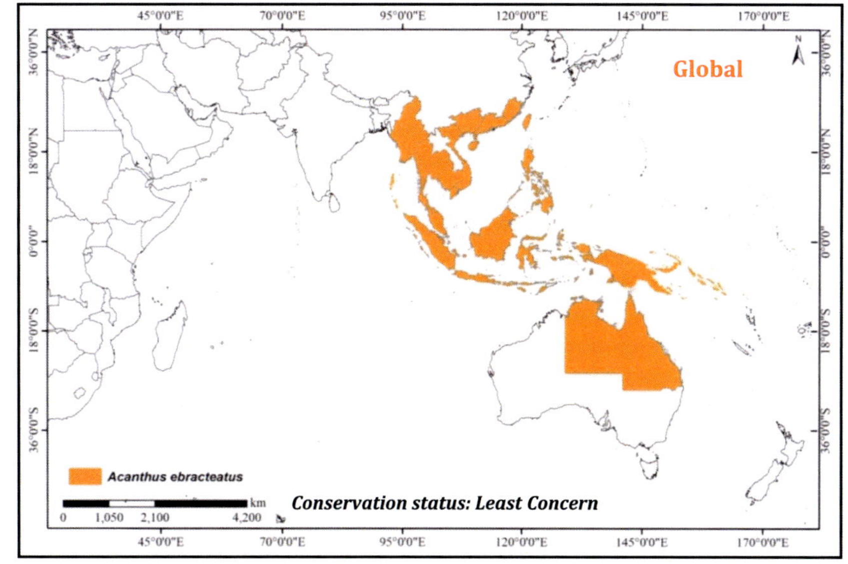

Habitat and Ecology

Found in landward edges of mangroves just above the high tide mark, and also occur in inner mangroves as understory. Individuals under shade are less serrated.

Phenology

Flowering and fruiting occur throughout the year.

Notes

The species status of *A. ebracteatus* is often doubted as the two species *viz., A. ilicifolius* and *A. ebracteatus* have similar vegetative characteristics. The difference between them is the presence or absence of bracteoles (which are often lost at anthesis), colour of petals (which is often variable) and size of flower and fruits. However, apart from these characters, the position of inflorescences and direction of stem axial spines at nodes aid the rapid differentiation of *A. ilicifolius* from *A. ebracteatus*. In *A. ebracteatus*, inflorescences are always terminal, and stem axial spines face downwards whereas in *A. ilicifolius* inflorescences are both terminal and axial and stem axial spines face upwards. Bract and bracteoles are always present and persistent in *A. ilicifolius* whereas in *A. ebracteatus* bracteoles are absent or minute and bracts are present, and both are lost before or at anthesis.

Fig. 5 Morphological features of *Acanthus ebracteatus*. (A) Habit; (B) prop roots; (C) stem axial spine facing downwards; (D) terminal inflorescences; (E) flower bud with only bract; (F) flower bud with bract and minute bracteoles; (G) style; (H) white corolla with purple stripe in middle of lower lip; (I) calyx lobes; (J) fruits; (K) removal of bract before anthesis; (L) mature fruit without bract and bracteoles; (M) small stamens with densely ciliated anther.

Acanthus volubilis Wall., Pl. Asiat. Rar. (Wallich). 2: 56, t. 172 (1831).

Acanthus volubilis is rare in India. It has restricted distribution in Sundarbans, Odisha and A & N Islands. It is easily distinguished from other *Acanthus* species by its unarmed leaves and stem, twining nature and terminal inflorescences with white flowers. This species is often confused with spineless form of *A. ilicifolius.*

Spicate inflorescence with white flowers

KEY CHARACTERS

- Twining shrub
- Leaves light green, elliptic, without spines
- Stem smooth twine like, stem axial spines absent
- Inflorescence terminal, spicate
- Bract single not persistent; bracteoles absent
- Flowers white in colour
- Mature Fruit 4 seeded capsule, ovoid, green.

Spineless leaves

white petals

Capsule like fruits

Stamen with hairy anther

Acanthus volubilis Wall

Taxonomy

Twining shrub: length 2–4 m (A). *Stem*: thin, green, branched, smooth, stem axial spines absent (I). *Roots*: occasionally above ground or prop roots on lower parts of reclining stem (G). *Leaves*: simple, opposite, without spines, succulent, elliptic or oblong lanceolate, leaf tip acute to obtuse with spiny edge, leaf base attenuate, margin entire without spines and dentate (B), 5–10 × 2.5–4 cm, ratio of length to width greater than 2; petiole short 0.5–2 cm long, green. *Inflorescence*: always terminal (C), 8–12 cm long; *Mature Flower bud*: ellipsoidal, 2–2.8 cm long; bract single not persistent; bracteoles absent (H); calyx 1–1.3 cm long, four lobed, upper lobe 1.3 × 0.5 cm, lower lobe 1 × 0.4 cm, enclosing the flower bud, lateral lobes narrow, 0.5–0.7 cm long, enclosed by upper and lower lobes (J); corolla white, 1.5–2 × 2–2.5 cm (D); stamens 4, 1.5 cm long, sub equal with thick hairy connectives; anther medifixed each with 2 cells aggregated around the style (E); ovary bilocular with 2 superimposed ovules in each loculus; style enclosed by stamens, 1–1.5cm long, capitate to pointed stigma exposed (F). *Mature Fruit*: 4 seeded capsule, ovoid, green, shiny, smooth, 2–2.5 × 1cm (K).

Fig. 6 Morphological features of *Acanthus volubilis*. (A) Habit; (B) leaves; (C) inflorescences; (D) white flowers; (E) stamens; (F) style; (G) stilt root; (H) mature bud with prominent bract; (I) delicate stem; (J) calyx lobes; (K) fruits.

Distribution

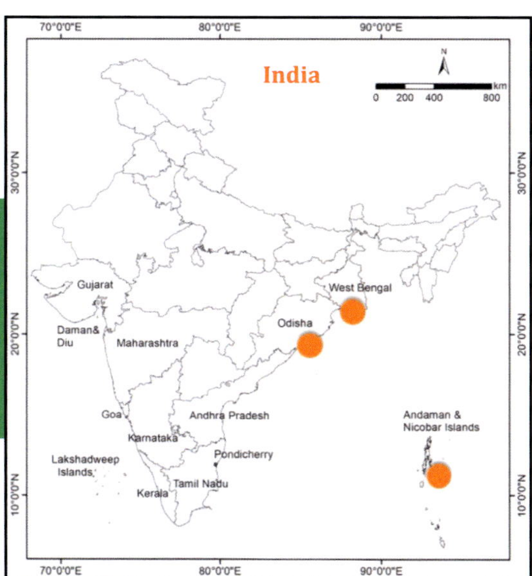

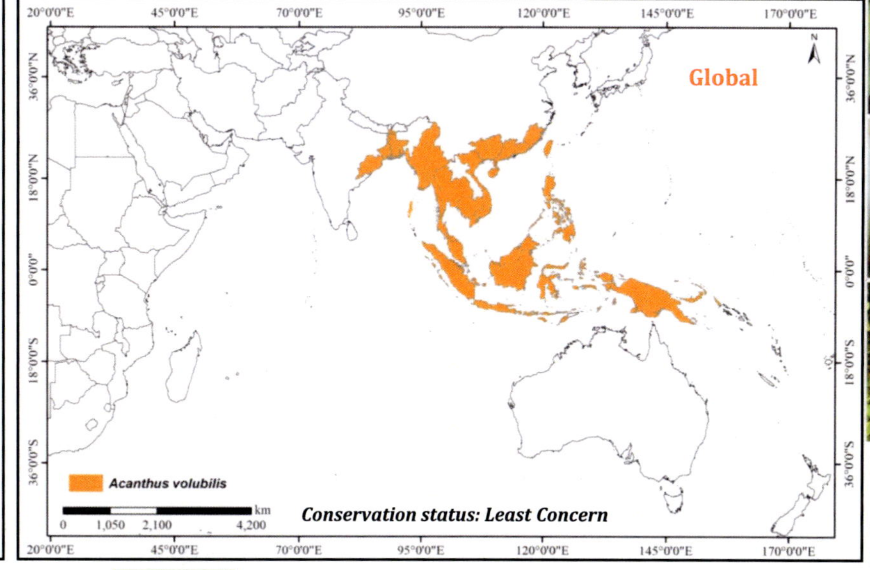

Habitat and Ecology

Often found in landward edges of mangroves just above the high tide mark, and also occur in inner mangroves as understory.

Phenology

Flowering March to April; Fruiting May to June.

Notes

The identity of *A. volubilis* is often confused with spineless/un-serrated form of *A. ilicifolius* (E.g. Dam Roy et al. 2009). However, *A. volubilis* can be easily distinguished from *A. Ilicifolius* and *A. ebracteatus* by its unarmed and twining, delicate sprawling stems, un-serrated elliptical leaves, white corolla and absence of bracteoles. In Indian literature *A. volubilis* is often included as mangrove associate. But it has considerable adaptation to colonize in mangrove niche viz., salt secreting leaves, water dispersed seeds and stilt root. A good number of *A. volubilis* populations occurs in Shoalbay Creek, Jirkatang and Kalighat Creek of Andaman Islands.

Population of *Acanthus ilicifolius* with spineless leaves

Genus: *Acrostichum*

Pteridaceae

Two species, viz., *Acrostichum aureum* and *A. speciosum*, representing Pteridaceae, are found in mangroves of India, of which *A. aureum* is common, while *A. speciosum* is present only in Odisha and Andaman Islands. It is important to note that unnamed putative hybrid between *A. speciosum* and *A. aureum* also known from Shoalbay Creek, Andaman Islands based on morphological intermediacy (Ragavan et al. 2013). *Acrostichum* species are rare, often dominant in mid and higher intertidal mangrove habitat. They are indicator species for mangrove degradation and also acidic soil condition.

Key to *Acrostichum* species

1. Leaflets narrowly acuminate at apex 2
1. Leaflets leathery, obtuse to obcordate and shortly mucronate at apex ...... ***Acrostichum aureum***
2. Leaflets leathery, leaf size more than 1 m, leaflet stalk > 1 cm, stipe up to 1 m ..... ***Acrostichum*** hybrid
2. Leaflets papery, leaf size not more than 1 m, leaflet stalk < 1 cm, stipe < 50 cm ...... ***Acrostichum speciosum***

Acrostichum aureum

Blunt leaf tip of *A. aureum*

Acrostichum speciosum

Elongated pointed leaf tip of *A. speciosum*

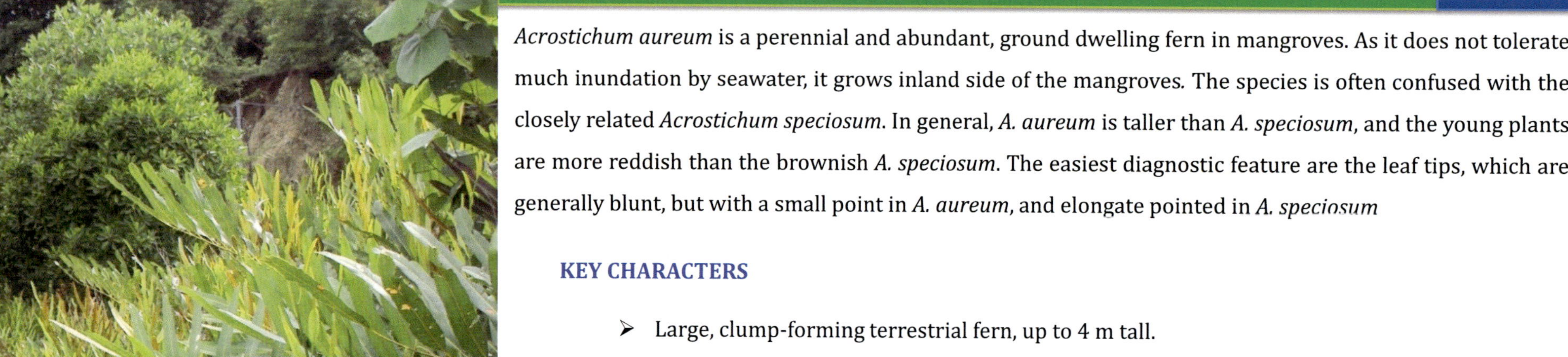

Acrostichum aureum L., Sp. Pl. 2: 1069 (1753).

Pteridaceae

Acrostichum aureum is a perennial and abundant, ground dwelling fern in mangroves. As it does not tolerate much inundation by seawater, it grows inland side of the mangroves. The species is often confused with the closely related *Acrostichum speciosum*. In general, *A. aureum* is taller than *A. speciosum*, and the young plants are more reddish than the brownish *A. speciosum*. The easiest diagnostic feature are the leaf tips, which are generally blunt, but with a small point in *A. aureum*, and elongate pointed in *A. speciosum*

KEY CHARACTERS

- Large, clump-forming terrestrial fern, up to 4 m tall.
- Stems are stout and erect, covered with large scales.
- Leaves are tall up to 3m length and but do not have more than 30 leaflets
- Fertile leaves are rusty-brown coloured, later turning dark brown and covered with large sporangia.
- Sterile leaflets are abruptly rounded or blunt, with a short tip.
- Spores are large and have a tetrahedral shape

Brownish young fertile leaves

Underside of the leaves with sporangia

Species feature

Blunt leaf tip

Acrostichum aureum L.

Taxonomy

Fern: height to 3 m (A). *Rhizome*: erect, scales dark brown to black (B). *Leaves*: unipinnate, 1-2.5 m long, up to 20 leaflets (A); stipe straw-coloured, > 0.5 cm in diameter, up to 1 m long. *Leaflet*: green, coriaceous, linear to oblong, apex obtuse to obcordate and shortly mucronate (C, D, E), base cuneate or rounded and irregular (F, G), margin entire, venation reticulate, 10–30 × 2-5 cm; leaflet stalk 1-3 cm long (F); costae strongly raised abaxially and grooved adaxially; fertile leaflet distal and smaller (H), 10–20 × 2-3.5 cm, rusty-brown coloured, later turning dark brown, underside covered with large sporangia, tip obtuse (I).

Distribution

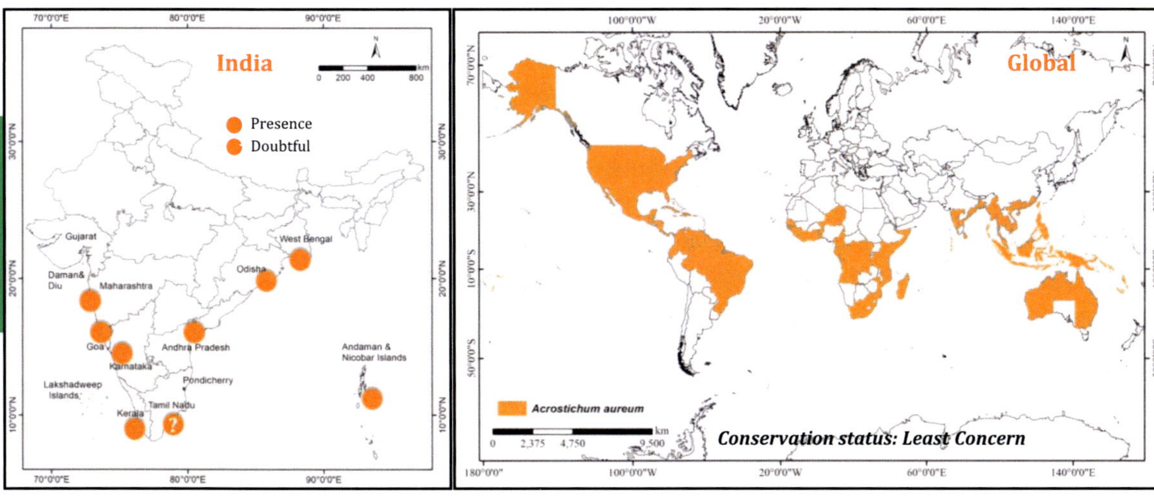

Morphological features of *Acrostichum aureum*. (A) Habit; (B) rhizome and stipe base with scales; (C) young sterile leaflets with pointed leaf tip; (D) blunt ends of mature sterile leaflets with mucronate; (E) obcordate leaf tip; (F) irregular base of leaflets; (G) long stalk; (H) young fertile leaflets with acute tip; (I) mature fertile leaflet with blunt end and sporangia underside.

Habitat and Ecology

Often found in inland side of mangroves and also found in inner mangroves as understory. It forms dense thickets in disturbed mangrove areas.

Phenology

It is a fern (pteridophytes) and so flowering and fruiting does not occur. Instead, spore bearing fertile leaves are formed throughout the year.

Notes

Acrostichum ferns are differentiated based on leaflet tip. *A. aureum* shows plasticity in shape of leaflet tip particularly young sterile and fertile leaflets that exhibit pointed tips like A. *speciosum* (Ragavan et al. 2013)

Acrostichum speciosum Willd., Sp. Pl., ed. 4 [Willdenow] 5: 117 (1810).

Pteridaceae

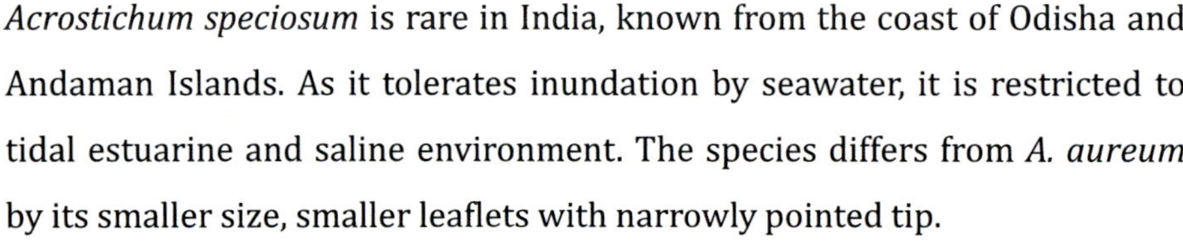

Acrostichum speciosum is rare in India, known from the coast of Odisha and Andaman Islands. As it tolerates inundation by seawater, it is restricted to tidal estuarine and saline environment. The species differs from *A. aureum* by its smaller size, smaller leaflets with narrowly pointed tip.

KEY CHARACTERS

- Small, clump-forming fern, up to 1.5m tall, rhizome erect, scales dark brown to black
- Leaves: unipinnate, not more than 1 m long, 10-16 leaflets.
- Leaflets green, papery, lanceolate with narrowly pointed tip and cuneate base, margin entire, venation reticulate.
- Spores are large and have a tetrahedral shape

Fertile leaf with sporangia

Species feature

Dark brown fertile leaf with sporangia

Elongated pointed leaf let tip

Single lanceolate leaf frond with alternate leaflets

Serrated margin of *Stenochlaena palustris*

Acrostichum speciosum Willd.

Taxonomy

Fern: height to 1.5 m (A). *Rhizome*: erect; scales dark brown to black (B); *Leaves*: unipinnate, not more than 1 m long, 10–16 leaflets (C); *stipe* straw-coloured, < 0.5 cm in diameter, 30–40 cm long (J). *Leaflet*: green, papery, lanceolate with narrowly pointed tip (D, E) and cuneate base (F), margin entire, venation reticulate, 10–20 × 1.5–2.5 cm; leaflet stalk < 1 cm (F); costae strongly raised abaxially and grooved adaxially; fertile leaflet distal, dark brown, covered with sporangia underside, 5–10 × 1–1.5 cm, tip pointed (G, H, I).

Distribution

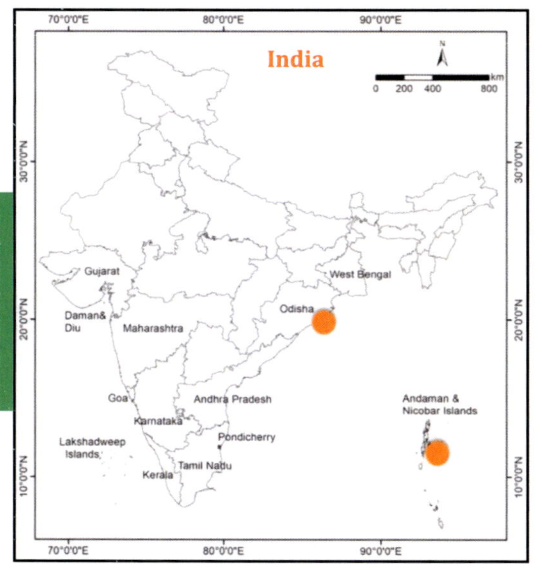

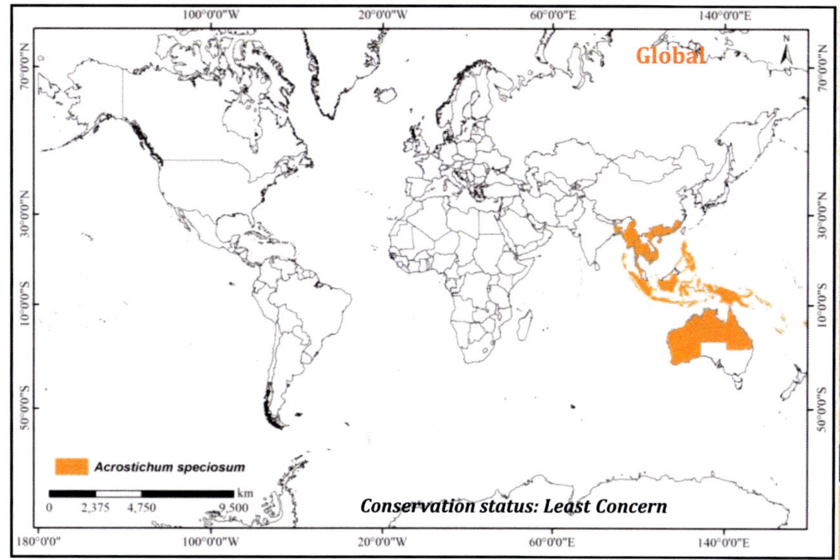

Morphological features of *Acrostichum speciosum*. (A) Habit; (B) rhizome and stipe base covered with scales; (C) sterile leaflets; (D & E) narrowly pointed tip of sterile leaflet; (F) small stalk and irregular base leaflets; (G) fertile leaflet with pointed tip; (H & I) mature fertile leaflet with sporangia underside; (J) comparison between stipe length of Acrostichum hybrid (left) and A. speciosum (right).

Habitat and Ecology

Occurs in mangroves that are more frequently inundated by tides and also present in association with *A. aureum* towards the landward edges of mangroves

Phenology

It is a fern (pteridophytes) and so flowering and fruiting does not occur. Instead, spore bearing fertile leaves are formed throughout the year.

Notes

The species is often confused with *Stenochlaena palustris* by the presences of pointed leaflets. Diagnostic characters of *A. speciosum* are leaflets with acuminate tip, entire margin of leaflets, and reticulate venation, while *Stenochlaena palustris* is climbing terrestrial fern with serrated margin and parallel venation of leaflets.

Acrostichum hybrid P. Ragavan, ISME electronic journal 12: 9-14 (2014).

Pteridaceae

This taxon is a putative hybrid between *A. aureum* and *A. specisoum*, described from Shoalbay creek of South Andaman Island by Ragavan et al. (2013). The hybrid status is supported by intermediate morphological characters (i.e. leaf length, leaflet length and width, stalk length, stipe length and fertile leaflet length) and distribution is limited to those areas where *A. speciosum* and *A. aureum* co-exist. The *Acrostichum* hybrid resembles *A. aureum* by its leathery leaflets and plant height; resembles *A. speciosum* by its pointed leaflets.

Fertile leaflet tip

KEY CHARACTERS

- Fern > 1.5m in height
- Rhizome erect; scales dark brown to black, broadly lanceolate
- Leaves unipinnate, > 1m L, 10–20 leaflets; stipe straw-colored
- Leaflets (pinnae) green, coriaceous, lanceolate with pointed tip base cuneate or rounded and irregular, margin entire, venation reticulate
- Fertile leaflets covered with sporangia underside, rusty brown, tip pointed

Dark brown scales

Habit

Sterile leaflet with acuminate tip

Sterile leaflet with acuminate tip

Fertile leaflets

Sporangia

Taxonomy

Fern: Height not more than 1.5 m (A). *Rhizome*: erect, scales dark brown to black, broadly lanceolate (C); *Leaves*: unipinnate, > 1 m long, 10–20 leaflets (G); stipe straw-coloured, > 0.5 cm in diameter, up to 1 m long (B); leaflet green, coriaceous, lanceolate with pointed tip (D, E), base cuneate or rounded and irregular (F), margin entire, venation reticulate, 10–30 × 2–5 cm; leaflet stalk 1–2 cm long (F); costae strongly raised abaxially and grooved adaxially; fertile leaflet covered with sporangia underside, rusty brown, 10–20 × 2-3.5 cm, tip pointed (H, I,J).

Distribution in India

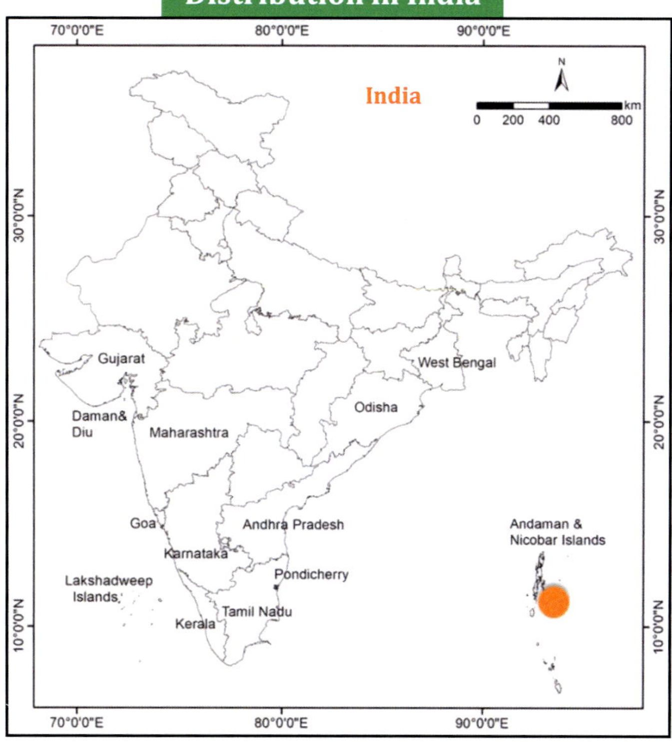

Habitat and Ecology

Found in inland side of mangroves where *Acrostichum aureum* and *A. speciosum* coexist.

Phenology

It is a fern (pteridophytes) and so flowering and fruiting does not occur. Instead, spore bearing fertile leaves are formed throughout the year.

Distribution

Acrostichum hybrids are known only from Andaman Islands and China

Notes

In the field, *Acrostichum* hybrid can be differentiated from *A. aureum* by its pointed tip of both sterile and fertile leaflets; and, from *A. speciosum* by its long stalk, long stipe and leathery leaves. Rhizomes are underground in both *A. speciosum* and the *Acrostichum* hybrid whereas in *A. aureum*, they are erect and aboveground. Zhang et al. (2013) reported natural hybridization between *A. aureum* and *A. speciosum* for the first time from China based on DNA sequences of three nuclear genes. Plastid DNA sequence data indicate that *A. speciosum* is most likely the maternal parent of this hybrid.

Acrostichum hybrid P. Ragavan

Morphological features of *Acrostichum* hybrid. (A) Habit; (B) stipe; (C) stipe base covered with scales; (D &E) mature fertile frond with narrowly pointed tip; (F) leaflet base and long stalk; (G) Fertile fronds; (H) Fertile fronds with long stalk and underside covered with sporangia; (I) pointed tip of fertile frond; (J) Mature fertile fronds.

Aegialitis rotundifolia Roxb., Pl. Ind. ii. 111 (1824).

Plumbaginaceae

The genus *Aegialitis* represents two species in global mangroves viz. *A. rotundifolia* and *A. annulata*. The two species are said to be distinguishable in pollen morphology. *Aegialitis rotundifolia* is restricted to coastal Bay of Bengal and Andaman Sea, whereas *A. annulata* is present in the coastal northern Australia and east Malesia. In the field, *A. rotundifolia* is easily identified by its ovate leaves with very long petiole. It is restricted to Sundarbans, Odisha and Andhra Pradesh in India.

Ovate leaf

KEY CHARACTERS

- Shrub or medium sized tree, grows 3-5m high
- Aerial root not prominent, dark, smooth, fissured or flaking with age.
- Leaves spirally arranged, ovate, petiole very long
- Inflorescence terminal, irregular, one-sided cymes with pairs of opposite linear bracteoles
- Calyx 5, green, united at base, petals 5, white, united at base, stamens 5, inserted on corolla tube, style 5, bilobed, seated on ovary
- Propagule pencil like, green, slightly curved, tip blunt, often upright in position.

Leaves with long petiole

Inflorescences with white flower

Upright propagules

Taxonomy

Shrub or Tree: grows up to 3–6m. *Bark*: dark grey or brown, smooth, fissured or flaking with age. *Roots*: no above ground roots. *Leaves*: spirally arranged, broadly ovate, green, 6–8 × 2–5 cm, glabrous, margin entire, apex rounded or bluntly acuminate; petiole very long, grooved adaxially and extended basally into a tubular leaf sheath with a completely encircling insertion. *Inflorescences*: terminal, irregular, one-sided cymes with pairs of opposite linear bracteoles, 4–10 flowered. *Mature flower*: pedicellate, small, 1.8 × 1.1 cm,, white, cup-shaped; calyx 5 lobed, small, 0.9 cm long, green, united basally and free apically; petals 5, alternate to calyx lobes, 1.2 cm long, white, free above with bluntly rounded lobes, and fused basally to form a corolla tube; stamens 5, 1.2 cm long, free, and inserted on the corolla tube alternately with the petal, filaments 0.7 cm long, whitish, glabrous, anthers bilobed, 0.2 cm long; style 5, 1.0 cm long, ovary is superior, one-chambered with a single basally attached anatropous ovule. *Mature fruit/propagule*: propagule pencil like, 4–10 m long, green, slightly curved, tip blunt, often upright in position.

Distribution

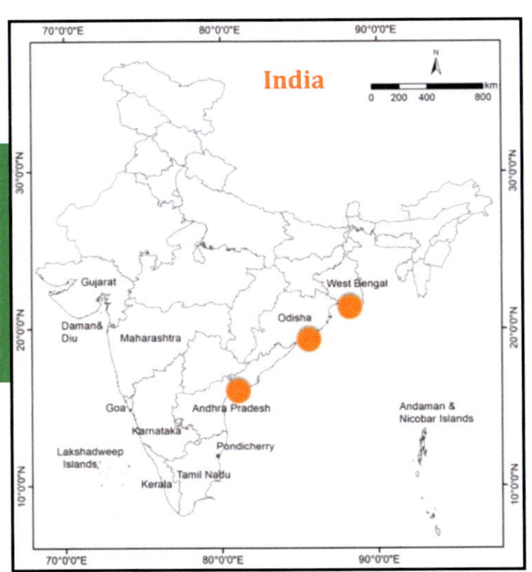

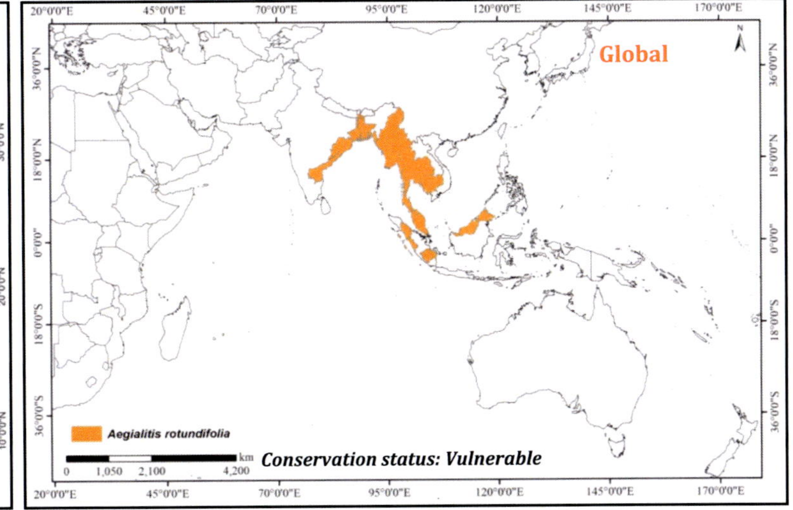

Habitat and Ecology

Often found in inner mangroves at mid estuarine position along with *Ceriops decandra, C. tagal, Bruguiera gymnorhiza and Excoecaria agallocha.* Also present at landward margin of mangroves inundated during normal high tides.

Phenology

Flowering February to April; Fruiting April to June.

Notes

Reports of *A. rotundifolia* from A & N Islands are erroneous (e.g. Dagar et al. 1991; Sreelakshmi et al. 2020a)

Aegiceras corniculatum Blanco, Fl. Filip. [F.M. Blanco] 79 (1837).

Myrsinaceae

The genus *Aegiceras* belonging to the family Myrsinaceae is represented by two species, viz., *Aegiceras corniculatum* and *A. floridum*, in the mangrove communities, of which only *A. corniculatum* is found commonly in Indian mangroves. In the field it is easily distinguished by its umbrella shaped flower cluster and distinctly curved, horn like fruits, while *A. floridum* have large straight fruits. *A. corniculatum* most commonly occurs along the landward margins of mangroves that are inundated by normal high tides, and fringes of seasonally brackish waterways.

Grey color bark with small lenticels

KEY CHARACTERS

- Shrub growing at sheltered inter tidal area, tolerant to high saline areas
- Aerial root not prominent, Bark smooth, reddish brown
- Leaves simple, alternate, leathery obovate to elliptical, leaf tip notched, cuneate at base, salt gland present, often pink red petiole
- Flowers white, terminal on long red peduncle, calyx -5 lobed green, corolla -5, partly fused, style protruding before flowers open
- Fruits yellow or pinkish, curved (like a small banana)

Simple stem base

Species feature

Glossy green obovate leaf

Spreading umbel off flower buds

White flower with abundant nectar

Cluster of horn shaped fruits

Aegiceras corniculatum Blanco

Taxonomy

Tree: medium sized, spreading, multi-stemmed, height to 5 m, evergreen (A). *Bark*: smooth, dark brown with small lenticels (E). *Roots:* not above ground (L). *Leaves*: simple, alternate, margin entire, dull green obovate to oblong obovate (B, C), apex obtuse, slightly emarginate, base cuneate, 5.5–9 × 2.5–5 cm, ratio of length to width c. 1.8; petiole terete, 1–1.5 × 0.2 cm. *Inflorescence*: terminal and axillary, clustered (B,F). *Mature Flowers bud*: ellipsoidal (I), 1–1.5 × 0.3–0.5 cm; peduncle 1.5 × 0.1cm; calyx lobes 5, green, 0.5 cm long (D); corolla lobes pointed, 5 lobed, white, 0.5 cm long, reflexed on maturity (G), apex pointed, fused at base and forms short tube of 1cm long; stamens 5, 0.3 cm long; ovary 0.5 cm long, conical extended to long simple style beyond the corolla tube (H); style 0.5–0.8 cm long. *Mature fruits*: curved, 4–8 × 0.3–0.5 cm, horn shaped, pointed at apex (J, K), calyx persistent and germination crypto-vivipary..

Distribution

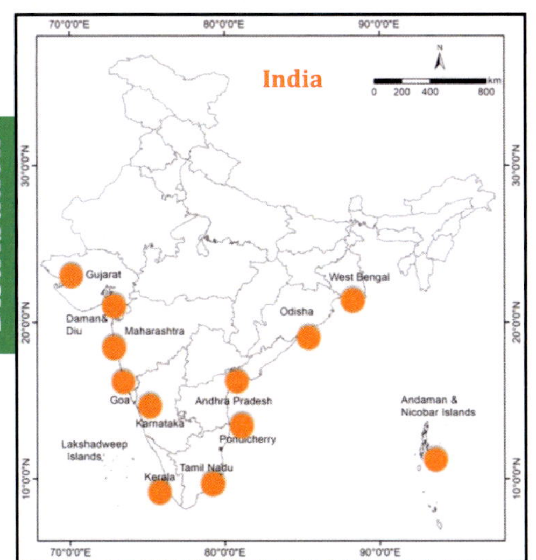

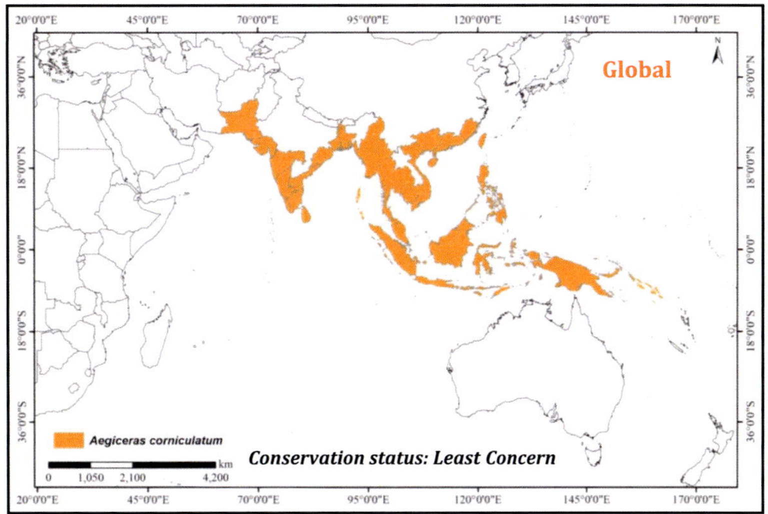

Habitat and Ecology

Often found in inner mangroves along with *Bruguiera* spp., *Ceriops* spp., and *Xylocarpus* spp. Also present at landward margin of mangroves inundated during normal high tides and fringing the banks at upstream region.

Phenology

Flowering and Fruiting throughout the year apparently. Flowering peaks at the month of March and April.

Notes

Aegiceras corniculatum is commonly called as 'mangrove banana' as the ripened fruits are morphologically similar to banana. In IWP, *A. corniculatum* is often referred to as black mangroves. Whereas in AEP *Avicennia* species are referred to black mangroves.

Morphological features of *Aegiceras corniculatum*. (A) Habit; (B) leafy rosette with young buds; (C) leaves; (D) buds; (E) bark; (F) flowers; (G) stamens; (H) style; (I) mature bud; (J&K) fruits; (L) simple stem base.

Aglaia cucullata (Roxb.) Pellegr., Fl. G?n. Indo-Chine 1 (1911). Meliaceae

Aglaia cucullata is the only representative of the genus *Aglaia* in global mangroves. Ecology and distribution of this species is not well known. Thus, this taxon is categorised as "Data Deficient". In India the distribution of *A. cucullata* is restricted to Sundarbans and Odisha. In the field *A. cucullata* is easily identified by the prominent pouch at the base of the terminal leaflet and prominent peg like pneumatophores with blunt tip.

Inflorescences

KEY CHARACTERS

- Tree grows up to 10m in height, prominent peg like pneumatophores present
- Bark pinkish grey or pale orange brown, slightly fissured
- Leaves compound, 3–4 pairs of leaflets, leaflets are sickle shaped, prominent pouch present at the base of terminal leaflets
- Inflorescences axial, multiflowered
- Flowers small, sub-globose, petals 3, yellow in colour
- Fruits are rounded, splits into three parts, each with one seed

Yellowish flowers

Leaf with 4 pairs of leaflets

Terminal leaflet with pouch at base

Pneumatophores

Bark

Fruit

Aglaia cucullata (Roxb.) Pellegr.

Taxonomy

Tree: spreading, grows up to 20m high, branched. *Bark*: pinkish grey or pale orange-brown, smooth, sometimes slightly fissured and flaky. Roots: stem base buttresses and peg like pneumatophores present. Leaves: imparipinnate, 3–4 pairs of leaflets, terminal leaflets possess prominent pouch at the base, leaflets are sickle shaped, petioles 10–15cm long. Inflorescences: axillary, multiflowered, 20–30cm long. Mature flower bud: small, sub-globose, 0.2–0.3cm long; calyx minute, cub shaped, green coloured; petals 3, yellowish, obovate, glabrous, 0.2–0.3 × 0.1–0.2 cm; staminal tube shorter than corolla, obovoid, six lobed, anthers 6, minute, protruding slightly through staminal tube aperture; ovary depressed-globose, loculi 3; stigma ellipsoid with 3 apical lobes and 6 longitudinal ridges; ovary and stigma together half of the length of staminal tube. Mature fruit: yellowish green, rounded, 5–7 × 3–4 cm, pericarp leathery, loculi 3, each contain single seed. Seeds: reddish brown, yellow or white aril covering half of the seed, 4–5 × 2–3 cm.

Distribution

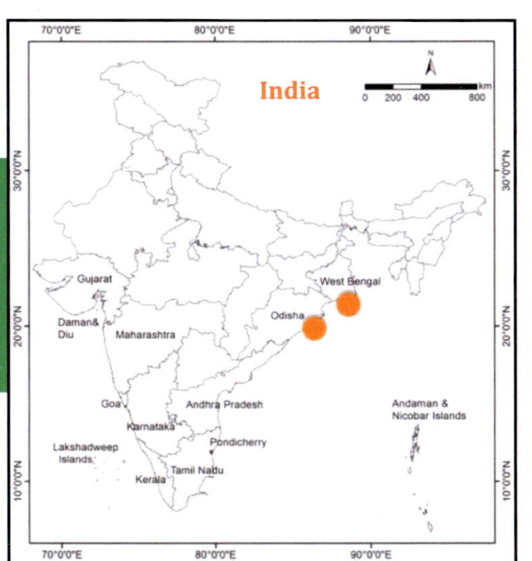

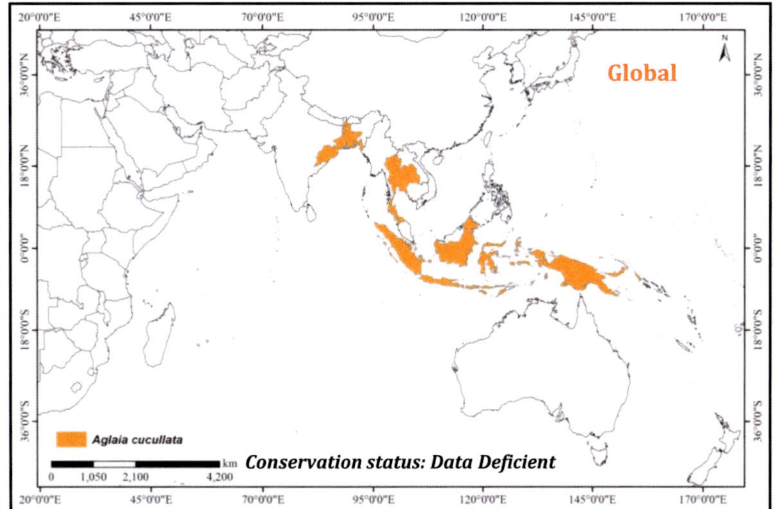

Habitat and Ecology

Often found upstream or mid-estuarine position in association with *Heritiera fomes*, *Xylocarpous moluccensis* and also fringing the banks at upstream regions.

Phenology

Flowering October to November; Fruiting December –January.

Notes

Earlier reports of *A. cucullata* from A & N Islands are erroneous (e.g. Dagar et al. 1991; Debnath 2004; Mandal and Naskar 2008). In most of the earlier literatures it is often classified as mangrove associate due to lack of adequate understanding on taxonomy and ecology. However, it has considerable adaptive features (like pneumatophores,) to colonize in mangrove niche.

Morphological features of *Aglaia cucullata*. (A) Habit; (B) leafy rosette (C) leaflets; (D) Bark (E) terminal leaflet with pouch at base; (F) Inflorescences; (G) pneumatophores; (H) yellowish flowers (I) rounded fruit.

Stunted population of *Avicennia marina* at Mumbai coast

Genus: *Avicennia*

The genus *Avicennia* L. belongs to the family *Acanthaceae* and consists of eight species worldwide-five in IWP viz., *A. marina, A. officinalis, A. alba, A. integra* and *A. rumphiana* and three in AEP viz., *A. germinans, A. schaueriana* and *A. bicolor*. In India three species viz., *A. alba, A. officinalis* and *A. marina* are known. In the field *Avicennia* species are easily recognized by its pencil like slender pneumatophores and pale grey green colour of underside of the leaves.

Pale grey green color of underside of the leaf

Key to *Avicennia* species

1. Inflorescences capitate 2
   Inflorescences spicate................3

2a. Corolla width < 0.5cm, style does not exceed the corolla, leaf apex acute.........***A. marina***

2b. Corolla width > 0.5cm, style exceed the corolla, leaf apex rounded..........***A. officinalis***

3. Corolla width <0.5 cm, style does not exceed the corolla, leaf apex acuminate....***A. alba***

Flowers of *A. officinalis* with prominent style

Flowers of *A. marina* with small style

Spicate Inflorescences of *Avicennia alba*

Pencil like pneumatophores

Avicennia alba Blume, Bijdr. Fl. Ned. Ind. 14: 821 (1826).

Acanthaceae

In India *Avicennia alba* is restricted to East coast of Mainland India. The key distinguishing features of *A. alba* is spicate inflorescences and fruits with highly prominent and elongated beak. *A. alba* also has ability to grow in a wide range of salinity. *Avicennia alba* is commonly distributed in Sundarbans, Bhitarkanika & Mahanadi delta of Odisha and Andhra Pradesh.

KEY CHARACTERS

- Large sized tree, grows up to 20m height
- Pencil like pneumatophores present
- Bark smooth, grey white to green, sometimes flaky
- Leaves are lanceolate with acuminate tip, often drooping
- Inflorescences spicate, flowers are small pale orange in color
- Fruits are capsule like, elongated with prominent beak, radicle glabrous

Spicate Inflorescences

Drooping leaves

Small flowers

Elongated fruits with prominent beak

Taxonomy

Tree: spreading, hight to 10–20 m (A). *Bark:* smooth, dark brown to black, slightly fissured (B). *Roots*: pencil like pneumatophores, 10–20 × 0.5–1 cm. *Leaves*: often drooping, lanceolate, apex acuminate, base attenuate, 5--10 × 2–5 cm, upper surface dark green (C), under surface dull pale and finely hairy; petiole green, 0.5–1.5 × 0.2 cm, glabrous above, hairy below. *Inflorescences*: terminal or subterminal, axillary, spicate, 3–7 bud pairs (D). *Mature flower buds*: small, ovate, 0.3–0.5 × 0.3 cm (E); bract single, narrowly triangular, apex acute; bracteoles two, lateral, triangular, edges ciliate, apices acute; calyx green, 4 lobed, lobes ovate, 0.3–0.4 × 0.2–0.4cm, edges ciliate or hairy, outer surface mostly pubescent; corolla 4 lobed, yellowish orange, 0.5 × 0.3 cm, apex acute, outer surface hairy and inner surface glabrous (F); stamens 4, alternate with corolla lobes, anther 0.1 cm long (F); style short, bilobed, 0.1 cm long, positioned below anthers (F); Ovary conical, around 0.2 cm long, upper portion densely tomentose, base glabrous. *Mature Fruit*: pod enclosing one propagule, elongate ellipsoid, 2–4 × 1–2cm, crypto-viviparous, distal tip sharply acute with narrow stylar beak (G), about 0.3-0.5 cm long; pericarp fleshy, outer surface slightly hairy, pale grey green, calyx persistent, 0.3–0.5 cm long; radicle 1 cm long, glabrous with short hairy collar.

Avicennia alba Blume

Morphological features of *Avicennia alba*. (A) Habit; (B) Bark; (C) Leaves; (D) Inflorescences; (E) Mature flower bud; (F) opened flower with calyx, style and stamens; (G) fruits with prominent elongated beak.

Distribution

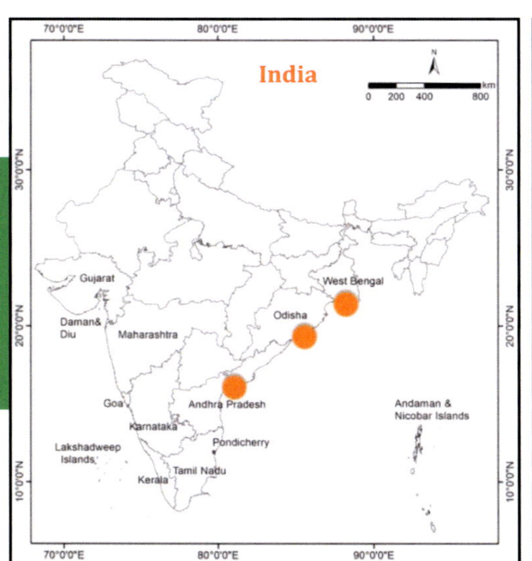

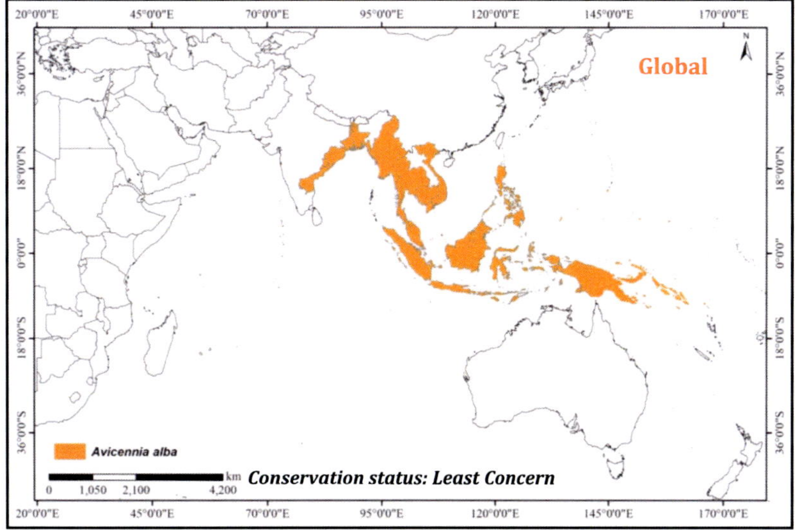

Habitat and Ecology

Often found in mid and downstream estuarine position. It can also grow in upstream areas where tidal influence is available.

Phenology

Flowering July to August; Fruiting September to October

Notes

Ecological variants of *A. marina* have been misidentified as *A. alba* in A & N Islands and West Coast (Dagar et al. 1991; Debnath et al. 2004; Sreelakshmi et al. 2018).

Avicennia marina (Forssk.) Vierh., Denkschr. Kaiserl. Akad. Wiss., Wien. Math.-Naturwiss. Kl. 71: 435 (1907).

Acanthaceae

Avicennia marina is the pioneer of mangrove species and widely distributed of all mangroves across the IWP region as well as in India. *Avicennia marina* has the ability to adapt to wide range of temperature, tidal inundation, salinity and substratum. These features provide this species to grow in sites where no other mangroves can survive. The species has a wide range of ecological variants among the population. *Avicennia marina* is easily identified by its small flowers, acute leaf apices, capitate inflorescence, glabrous radicle and rounded fruits with minute beak.

Mature fruits without beak

KEY CHARACTER

- Tree grows up to 10-15m high with prominent pencil like pneumatophores
- Leaves elliptic to lanceolate, highly variable based on edaphic variables
- Bark smooth, grey-white to green bark made up of thin, stiff, brittle flakes
- Inflorescences terminal as well as axial, capitate
- Flowers small, pale yellow, style does not exceed the corolla
- Fruits almond sized, dark green with obscure beak
- Radicle smooth

Glabrous radicle

Pale yellow color flowers

Capitate inflorescence

Leaves with acute leaf tip

Smooth bark

Avicennia marina (Forssk.) Vierh.

Taxonomy

Tree: spreading, height to 20 m (A). *Bark*: smooth, chalky white with thin flakes (R). *Roots*: pencil like pneumatophores, 10–25 × 0.5–1 cm (Q). *Leaves*: shapes widely variable, elliptic or broadly elliptic or lanceolate, apex variably pointed, acute to obtuse (B, C, D), base attenuate to cuneate, 4.5–10.5 × 2–4.5 cm, ratio of length to width is greater than 2, upper surface glossy green, under surface dull pale and finely hairy; petiole green, 0.5–1.5 × 0.2 cm, glabrous above and hairy below. *Inflorescence*: terminal or sub-terminal axillary, capitate, 2–5 buds pairs (E, F); *Mature flower buds*: ovate, 0.5–0.8 × 0.3 cm (G); bract single, ovate, margin hairy (K); bracteoles two, lateral, ovate, margin hairy (K); calyx green, 5 lobed, margin hairy, outer surface slightly hairy at base (J); corolla 4 lobed, yellowish orange (H), 0.5 × 0.3 cm, apex rounded, outer surface hairy and inner surface glabrous; stamens 4, alternate with corolla lobes (L), anther 0.1 cm long (I); style short, bilobed 0.1 cm long, positioned below anthers (I); Ovary conical, around 0.2 cm long, upper portion densely tomentose, base glabrous (J). *Mature Fruit*: pod enclosing one propagule, 1.5–3 × 1–2cm, crypto-viviparous, stylar beak persistent, variable 0.1 cm long (N, O); pericarp fleshy, outer surface slightly hairy, pale grey green, calyx persistent, 0.3–0.5 cm long (M); radicle 1 cm long, glabrous with short hairy collar (P).

Distribution

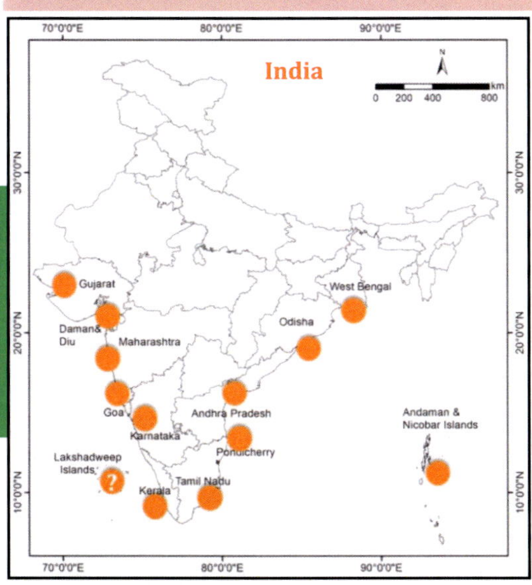

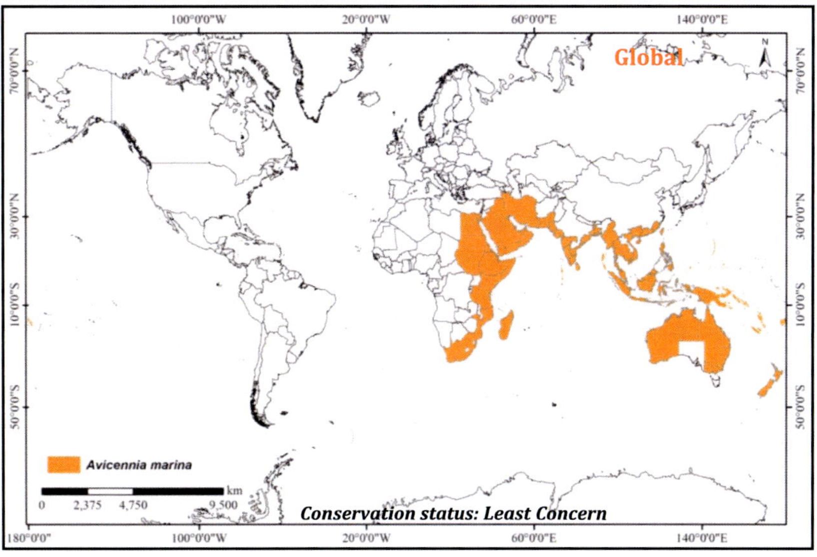

Habitat and Ecology

Often found in high intertidal and intermediate estuarine position, also present in downstream and low intertidal areas. It is a dominant species in highly organic polluted areas and arid and dry habitats. it is an abundant species along the east and west coast of Mainland India.

Phenology

Flowering March to May; Fruiting June to August.

Notes

Two varieties of *A. marina* are also known from India. *Avicennia marina var. acutissima* is identified based on sessile leaves, and *A. marina var. resinifera* based on very acute leaf tip. Their occurrence is reported along the coast of Maharashtra and Gujarat. Since *Avicennia marina* has a wide variation in leaf shape, peak of fruits, style length in relation to anther (Ragavan 2015), further studies are required to confirm the varietal differences among *A. marina* based on pubescence in bract, bracteoles and calyx lobes.

Morphological features of *Avicennia marina*. (A) Habit; (B) leaves with acute tip; (C) leaves with blunt tip; (D) lanceolate like leaf; (E &F) inflorescences; (G) mature bud; (H) hairy petals; (I) stamens; (J) style and calyx lobes with hairy margin; (K) bract and bracteoles; (L) blunt apex of petal; (M) fruits without beak; (N) fruits with prominent beak; (O) intermediate beak; (P) glabrous radicle hairy at collar;(Q) pneumatophores; (R) bark.

Avicennia officinalis L., Sp. Pl. 1: 110 (1753).

Acanthaceae

Avicennia officinalis is commonly distributed in India and distinguished from *A. marina* by its large sized flowers with relatively long style, rounded leaves, fruits with prominent beak and hairy radicle. Occurs on the landward margins of mangrove swamps, also found in upstream and mid estuarine position along tidal creek banks.

Inflorescences

KEY CHARACTERS

- Tree grows up to 10-15m height
- Pneumatophores slightly thicker than *A. marina*
- Bark smooth, grey white to green, slightly fissured and sometimes flaky
- Leaves ovate oblong or elliptic ovate
- Inflorescences capitate
- Flowers larger than *A. marina*, pale orange in color, style exceeds the corolla
- Fruits are capsule like with prominent beak
- Radicle hairy

Species feature

Leaf with rounded leaf tip

Immature bud with calyx having hairy margin

Large seized flowers with prominent long style

Fruits with hairy pericarp and prominent beak

Avicennia officinalis L

Taxonomy

Tree: spreading, height to 20 m (A). *Bark*: grey-brown, finely fissured (N). *Roots*: pneumatophores thicker than *Avicennia marina*, tip blunt, 20–30 × 0.5–0.1 cm (K). *Leaves*: obovate, oblong-obovate or elliptic-oblong, apex rounded (B, C), 5–11 × 3–7 cm, upper surface bright satiny green, under-surface pale finely pubescent; petiole 1-2 cm long, glabrous above, often pubescent below. *Inflorescences*: mostly capitate with 2–4 opposite, decussate bud-pairs (G). *Mature flower bud*: ovate, 0.8–1.2 cm long at anthesis (E); bract single, circular, edge hairy (D); bracteoles two, lateral, oblong, apices rounded, edges hairy (D); calyx lobes 4, ovate, 0.5 cm long, edges hairy, outer surface smooth (E); corolla lobes 4, yellow-orange, 0.3–0.5 × 0.5–0.8 cm, reflexed, apices rounded, outer surface hairy, inner surface glabrous (F); stamens 4, alternate with corolla lobes, 0.3 cm long, anther 0.1 cm long (J); style bilobed, 0.3 cm long, glabrous, not exceeding the anther, but exceeds corolla (H, I); ovary ampulla-shaped, about 0.4 cm long, densely tomentose. *Mature Fruit*: ellipsoid, 1.5–3.5 × 1–2 cm, distal tip acute, with narrow persistent stylar beak (M); pericarp outer surface pale grey-green, hairy, calyx persistent on pericarp, 0.5–0.8 cm long from base, 0.8–1.3 cm diameter; radicle about 1.3 cm long, densely hairy along the full length (L).

Distribution

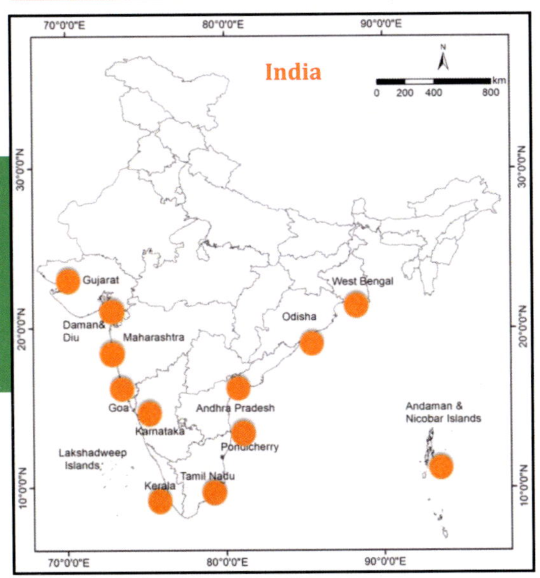

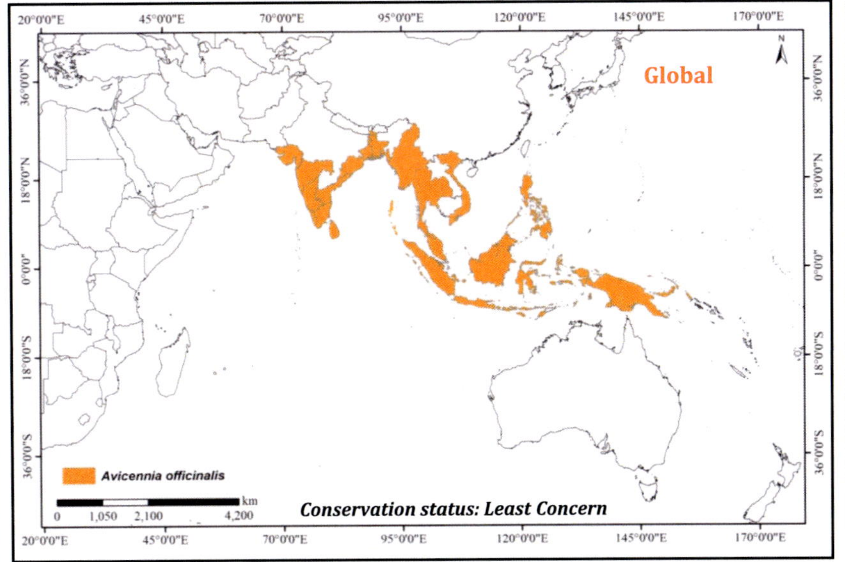

Morphological features of *Avicennia officinalis*. (A) Habit; (B) leafy rosette; (C) leaf with rounded apex; (D)Young bud; (E) mature bud with calyx hairy at margin; (F) long stamens; (G) inflorescences; (H&I) style; (L) long filament; (K) pneumatophores; (L) hairy radical; (M) pubescent fruits with prominent beak; (N) fine fissured bark.

Habitat and Ecology

Often found in low and high intertidal position and also occur in mid and upper estuarine position along the banks of the creek.

Phenology

Flowering June to July; Fruiting August to September

Notes

Avicennia officinalis has morphological similarity with *Avicennia integra* (species known from Australia). *A. integra* is distinguished by calyx edges that are entire rather than ciliate or hairy as in *A. officinalis*. Supportive characters are larger dimensions of the flower in *A. integra* compared with *A. officinalis*, particularly those of calyx width and length and length of anthers (Duke 1991).

Brownlowia tersa (L.) Kosterm., Monogr. Gen. Herit., etc. (Penerb. Madjel. Pengetah. Indones. i.) 73 (1959), inobs.

Malvaceae

The genus *Brownlowia* belongs to the family Malvaceae and is represented by two species viz., *Brownlowia tersa* and *B. argentata*, in mangrove communities. Of these only *B. tersa* is known from Indian mangroves and occur in Sundarbans, Odisha, Andhra Pradesh and A & N Islands. *Brownlowia tersa* can be readily recognized in the field by its brown-scaly twigs, lanceolate leaves with dull silvery under surface and pear-shaped, 2-valved fruits.

Species feature

Silvery grey underside

KEY CHARACTERS

- Shrub, usually 1.5-2 m tall
- Leaves are lanceolate, upper surface is glossy, smooth dark green, while the lower surface is grey-green and covered with a dense layer of tiny, hairy scales
- The flowers occur in axils or terminal, few-flowered,
- The calyx is bell-shaped, with 3–5 lobes
- The corolla is pink with a yellow base, slightly longer than the calyx
- Fruit is a woody capsule or nut, bi-lobed or heart-shaped and is pale greyish-green, covered with small, brown warts

Habit

Flower with numerous stamens and petals

Calyx

Bi-lobed fruit

Brownlowia tersa (L.) Kosterm.

Taxonomy

Shrub: spreading, height to 2 m (A). *Roots*: no above ground roots. *Branches*: grey and smooth, small branches often pubescent; *Leaves*: lanceolate (B), apex pointed, base rounded, 5–20 × 2–5 cm, ratio of length to width >2 , underside silvery grey (C), above glossy green in colour; petiole small, 1–2 cm long, grey in colour; *Inflorescences*: terminal or axial, few flowered, up to 4 cm long (G); *Mature flower bud*: rounded, 0.4–0.6 cm long (F); calyx, brown, 0.5 cm long, 3–5 lobed, bell shaped (E); petals 5, yellowish or pink in colour (D), 0.3 × 0.2 cm, apex rounded; stamens numerous with yellow colour anther (D); *Mature fruits*: woody capsule, heart shaped, up to 1.5 cm long, greyish green in colour, covered with small, brown warts (H), single seeded.

Distribution

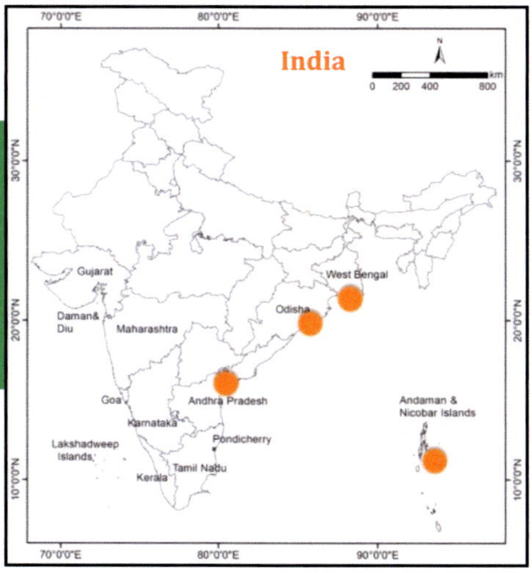

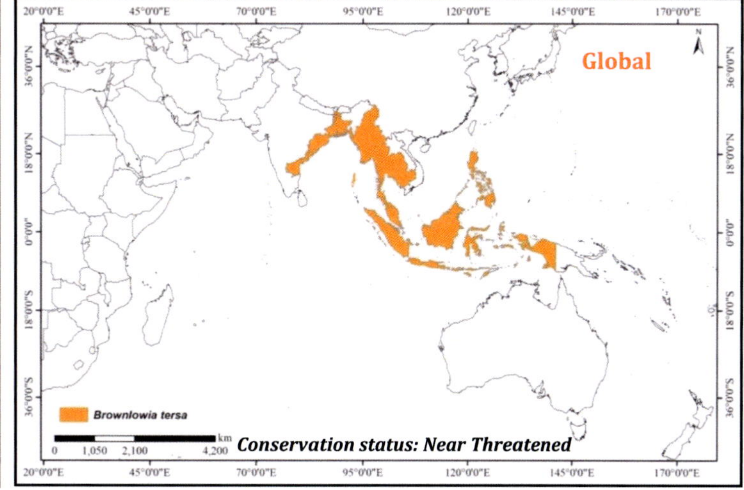

Habitat and Ecology

Found in banks of the upstream estuarine position and also observed in inner mangroves in association with *Bruguiera gymnorhiza* and *Avicennia officinalis*.

Phenology

Flowering February to March; Fruiting April to July.

Notes

Brownlowia tersa is "Near Threatened" mangrove species. In the field *B. tersa* is often misidentified as *Heritiera* species in absence of flowers and fruits due to similar leaf characteristics viz, silvery grey underside and dark green colour of upper side.

Morphological features of *Brownlowia tersa*. (A) Habit; (B) lanceolate leaves; (C) slivery grey underside of the leaf; (D) flower with numerous stamens and petals; (E) calyx; (F) flower buds; (G) inflorescences; (H) matured fruits.

Knee roots of *Bruguiera* spp.,

Knee roots

Genus: *Bruguiera*

Rhizophoraceae

Bruguiera is an IWP genus within the family Rhizophoraceae. *Bruguiera* species are broadly divided into two groups i.e. large and solitary flowered *Bruguiera* species (*B. exaristata, B. gymnorhiza, B. sexangula, B. × rhynchopetala* and *B. × dungarra*) and small and multi flowered Bruguiera species (*B. cylindrica, B. × hainesii and B. parviflora*). In India four species namely *B. gymnorhiza, B. sexangula, B. cylindrica* and *B. parviflora* are present. *Bruguiera* are distinguished from other members of Rhizophoraceae by its calyx with 8-16 pointed lobes, 16-32 stamens, and explosive pollen release.

Tall tree of ***Bruguiera gymnorhiza*** **(more than 30m height)**

Key to *Bruguiera* species

1. (a) Flowers multiple (2-5), small; petal spine exceeds lobes.................2.

1. (b) Flowers solitary, large; petal spine shorter than lobes.........3

2. (a) Fruit calyx ribbed, lobes adpressed, propagules straight thin with blunt tip, leaves yellowish green..***B. parviflora***

2. (b) Fruit calyx smooth, lobes reflexed, propagules slightly curved with acute tip, leaves dark green...***B. cylindrica***

3. (a) Petal bristles 3, greater than 2 mm, calyx colour variable (red, pink, green, white, yellowish green), calyx tube smooth........................... ***B. gymnorhiza***

(b) Petal bristles minute (1-2) or absent, calyx yellow, calyx lobe ridged.........***B. sexangula***

Single large solitary flower buds of *Bruguiera gymnorhiza*

Small multi-flowered racemes of *Bruguiera cylindrica*

Young, white bi-lobed petal with petal spine and petal bristles

Mature brown Petal

Bruguiera gymnorhiza (L.) Lam., Tabl. Encycl. 2(5.2): 517 (-518); 2(2.2): t. 397 (1819).

Bruguiera gymnorhiza is common in India. *Bruguiera gymnorhiza* is distinguished from other *Bruguiera* species by its large solitary-flowered inflorescence with petals having a spines slightly shorter than paired lobes as distinct from *B. parviflora*, *B. cylindrica* and *B. × hainesii*; and its acutely pointed petal lobes with 3-4 bristles being distinct from the more rounded petal lobes with lesser bristle number of *B. sexangula, B. ×dungarra* and B. *×rhynchopetala*.

Mature cigar shaped propagules

KEY CHARACTERS

- Tree grows up to 10–20m height, knee roots and buttresses present
- Bark dark brown and rough, slightly fissured and lenticels present
- Large leaves occur in clumps at the end of branches, elliptic
- Inflorescences solitary
- Calyx smooth, often red in colour but varies widely
- Petals bilobed, petal spine does not exceed the petal lobe, petal bristles 2–3, conspicuous
- Propagules are green and cigar shaped, 10–20 cm long

Species feature

Bright red calyx

Green color calyx

Petal spine > petal lobe

Mature flower with brown petals

Petal bristles at apex

Bruguiera gymnorhiza (L.) Lam.

Taxonomy

Tree: columnar or spreading, height to 30 m, evergreen, erect (A). *Bark*: dark grey to brown, rough, highly fissured (E). *Roots*: knee like pneumatophores (F), buttress at base of the stem (D). *Leaves*: simple, opposite, green, leathery, elliptic oblong, 7.5–15 × 3.5–7 cm, ratio to length to width c. 2, apex acute, base cuneate (B); petiole green, 1.5-5.5 × 0.4 cm. *Inflorescence*: axillary, 1-flowered (C); peduncle 1–2 × 0.2 cm. *Mature flower bud*: ellipsoidal, 3–3.5 × 1cm (H); calyx red or green or yellowish green or white, bright red on exposure of sunlight (C, H, I), smooth, grooved above lobe juncture, 10–14 lobes (G), apex acute, 1.5–2.5 cm long; petals 10–14, white, turns brown on maturity, 1-1.5 cm long, bilobed, apex acute with 2–3 bristles (rarely 4), 0.2–0.3 cm long, petal spine between the petal lobe is conspicuous (J), 0.5–0.8 cm long, based on the length of petal spine with respect to petal lobe three forms can be recognized i.e., petal spine equal to petal lobe, longer than petal lobe and shorter than petal lobe; stamens twice the number of petal, 20–28, each two enclosed by one petal, creamy white and turns brown on maturity, 0.5 cm long; style 1.5-2.5 cm long, tip 3–4 lobed (G). *Mature fruit*: present within the calyx tube. *Mature Hypocotyls*: green, cigar shaped with longitudinal ribs, 15–25 × 1–2 cm, tip blunt, calyx persistent (K); plumule 0.3 cm long.

Distribution

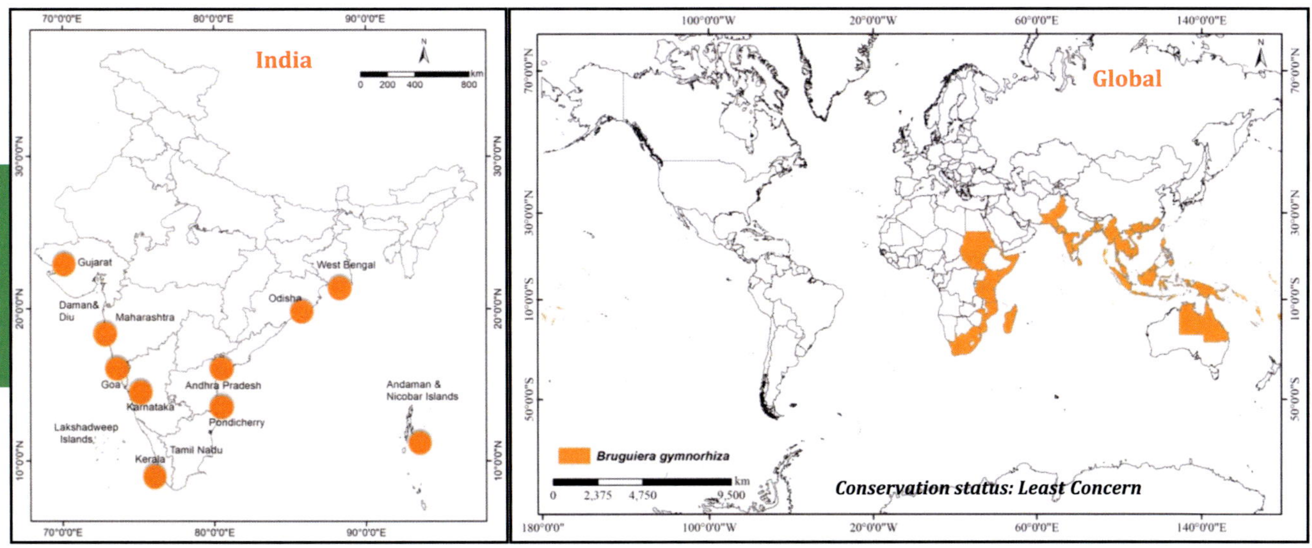

Habitat and Ecology

Often occurs in intermediate and upstream estuarine position.

Phenology

Flowering June to August; Fruiting September to January.

Notes

Bruguiera gymnorhiza is often identified based on red colour of calyx, but calyx colour is highly variable. Petal spine length in relation to petal lobe and number and length of petal bristles are reliable diagnostic features (Duke 2006).

Morphological features of *Bruguiera gymnorhiza*. (A) Habit; (B) leafy rosette with flowers; (C) reddish calyx tube; (D) buttresses; (E)bark; (F) knee roots; (G) flower; (H) mature bud; (I) bud with long peduncles; (J) bilobed petal with bristles and petal spine; (K) cigar shaped propagules with longitudinal ridges

Bruguiera parviflora (Roxb.) Wight & Arn. ex Griff., Trans. Med. Soc. Calcutta 8: 10 (1836).

Rhizophoraceae

Bruguiera parviflora is common in Andaman Islands, while it has restricted distribution in mainland India. *Bruguiera parviflora* is easily recognized in the field by its light green or yellowish green small leaves with slender petioles. It is distinguished from other *Bruguiera* species by its multi-flowered inflorescences, small flowers with narrow, long, ribbed calyx tube with short lobes slightly reflexed during flowering, and adpressed during propagule maturation and spaghetti-like mature propagules.

KEY CHARACTERS

- Tree grows upto 10m height, knee roots and buttresses present
- Bark rough, dark brown, slightly fissured
- Leaves are small yellowish green in color, clumps at the end of the branches, elliptic
- Inflorescence axial, 3–5 flowered
- Flowers small, calyx tube green, ridged, calyx lobes minute
- Petal spine exceeds the petal lobe, apex with 3 small petal bristles
- Propagules are pencil like, smooth, tip blunt, calyx tube persistent

Species feature

Yellowish green small leaves

Small multi-flowered inflorescences

Long persistent calyx tube

Pencil like long propagules

Bruguiera parviflora (Roxb.) Wight & Arn. ex Griff.

Taxonomy

Tree: columnar or spreading, height to 20 m evergreen (A) *Bark*: grey, smooth, slightly fissured on maturity (F). *Roots*: knee like pneumatophores (E), buttress at stem base. *Leaves*: simple, elliptic, opposite, margin entire, yellowish green (B), 6.5–11 × 2.5–4.5 cm, ratio of length to width >2, apex acute, base cuneate; petiole yellowish green, 1.5–2.5 × 0.2 cm. *Inflorescences*: axillary, 3–5 flowered (D); peduncle 1.5-2 cm long; *Mature flower bud*: oblong, 1–1.7 × 0.3 cm; calyx 8 lobed (C), 0.2 cm long, calyx tube 1–1.5cm long; petals 8, whitish orange, 0.1–0.2 cm long, bilobed, apex with 3 bristles (C), 0.2 cm long, petal spine exceeding lobes, 0.3 cm long; stamens 16, 2 enclosed in each petal, 0.2 cm long; style slender 0.2 cm long; *Mature Fruits*: present within calyx tube, germination viviparous. *Mature hypocotyls*: thin, pencil like, tip blunt, yellowish green (I), 10–20 × 0.5 cm, calyx persistent, lobes enveloping the hypocotyls (G); plumule 0.2 cm long.

Distribution

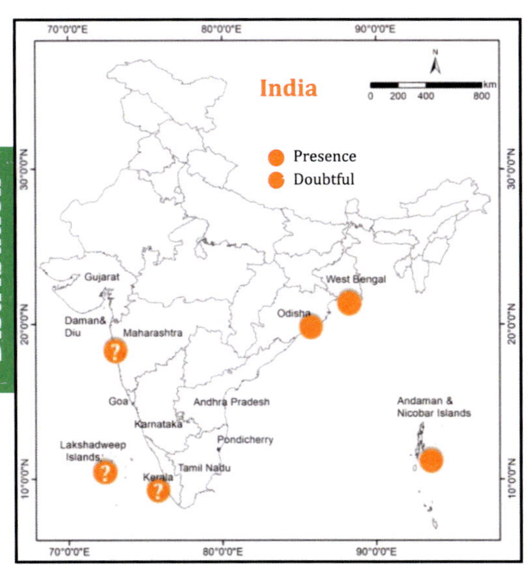

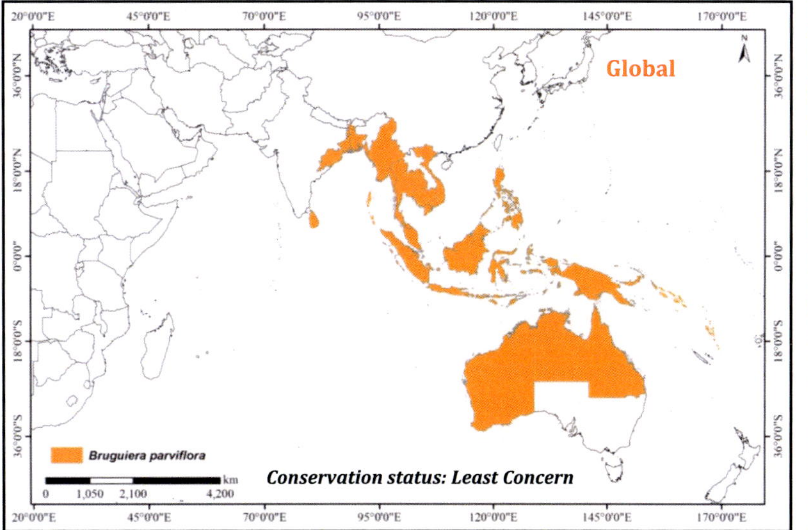

Habitat and Ecology

Often found in mid intertidal and intermediate estuarine position on soft muddy substratum.

Phenology

Flowering November to January; Fruiting February to April.

Notes

In the field *B. parviflora* is often confused with *B. cylindrica*. Distinguished from *B. cylindrica* from its yellow wider petiole, ribbed, narrow calyx tube, whereas in *B. cylindrica* has greenish petiole, and smooth calyx tube with reflexed lobes. Chavan (2013) reported the occurrence of *B. parviflora* from Maharashtra coast; however, its distribution along the west coast of mainland India needs to be verified.

Morphological features of *Bruguiera parviflora*. (A) Habit; (B) leafy rosette; (C) flower; (D) inflorescences; (E) knee roots; (F) bark; (G) ridged calyx tube with enveloped lobes; (H) young propagule; (I) mature propagule with blunt tip.

Bruguiera sexangula (Lour.) Poir., Encyc. [J. Lamarck & al.] Suppl. 4. 262 (1816).

Rhizophoraceae

Bruguiera sexangula is a rare mangrove species in India. Its distribution is restricted to Sundarbans, Odisha and Kerala. *Bruguiera sexangula* is easily identified in the field by its ridged calyx tubes, petals with petal spine shorter than petal lobe, petal apex without bristles or 1–2-minute bristles and propagules in slightly upright position.

Prominently ridged calyx tube

KEY CHARACTERS

- Tree grows up to 5–10m height, stem base buttressed, prominent knee roots present
- Bark greyish to pale brown, slightly fissured, lenticels present
- Leaves elliptic, dark green or yellowish green, petiole green
- Inflorescences axial, solitary
- Calyx yellowish or yellowish green, calyx tube ridged, calyx lobes 10–14
- Petals as many as calyx lobes, bilobed, petal spine shorter than petal lope, petal bristles absent or minute
- Propagules cigar shaped, slightly upright in position, smaller than *B. gymnorhiza*, not ridged

Upright young propagule

Dark green leaves

Yellowish flower bud

Petal with hairy margin

Petal spine shorter than petal lobe

Bruguiera sexangula (Lour.) Poir.

Taxonomy

Tree: columnar or spreading, height to 10 m evergreen (A). *Bark*: grey to pale brown, slightly fissured, plenty of lenticels present (B). *Roots*: stem base buttressed (C) and knee like pneumatophores; *Leaves*: simple, elliptic, opposite, margin entire, dark green (D) or yellowish green, 8–13 × 4–6 cm, ratio of length to width >2, apex acute, base cuneate; petiole green or yellowish green, 1.5-2.5 × 0.2 cm. *Inflorescences*: axillary, single flowered (E); peduncle 0.6–1.2 cm long; *Mature flower bud*: elliptic, 2.5–4 × 0.5–1.5 cm; calyx yellowish (F), 10–14 lobed, 1.5-2 cm long, calyx lobe margin grooved (E), calyx tube 1–1.5cm long, prominently ridged (F); petals as many as calyx lobes (G), whitish when young turns brown on maturity, 1.5–2 cm long, bilobed, petal spine not exceeds the petal lobe (H), peal apex without bristles or with 1-2 minute bristles of 0.2 –0.4 cm long; stamens double the number of petals, 2 enclosed in each petal, 0.5 cm long; style 1–2 cm long, tip 3–4 lobed (G). *Mature fruit*: present within the calyx tube. *Mature Hypocotyls*: green, cigar shaped without longitudinal ribs (I), 10–15 × 1–1.5 cm, tip blunt, calyx persistent; plumule 0.3 cm long.

Distribution

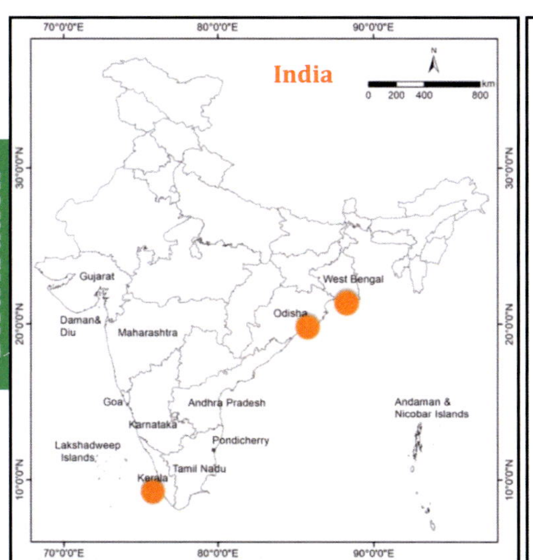

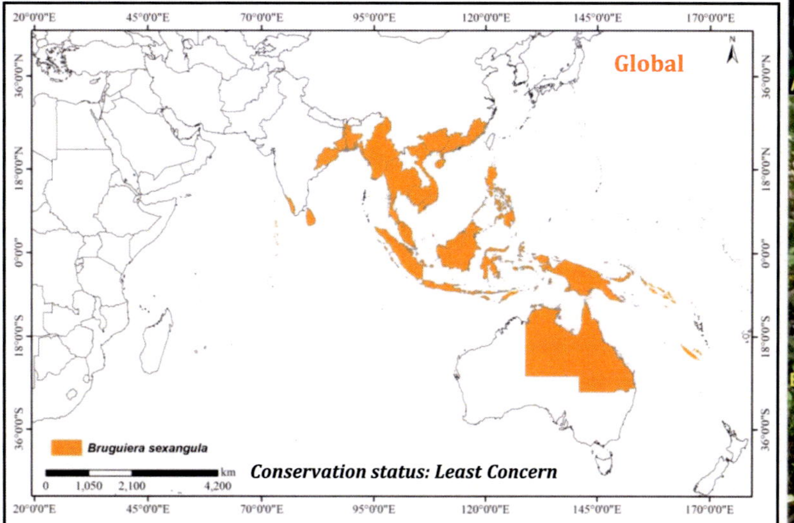

Habitat and Ecology

Often found in mid intertidal and intermediate estuarine position on soft muddy substratum

Phenology

Flowering and Fruiting occurs through the year. Flowering peaks July to August, fruiting from September to November

Notes

This taxon is distinguished from *B. cylindrica* and *B. parviflora* by its solitary inflorescences, but often difficult to differentiate from *B. gymnorhiza*. *B. sexangula* is often distinguished from *B. gymnorhiza* by its yellowish colour of the calyx, but calyx colour is highly variable in both the species. Thus, length of the petal spine in relation to petal lobe, number and length of the petal bristles are highly useful to ensure the correct identity of *B. sexangula*. The reports of *B. sexangula* from Andaman Islands are erroneous (e.g. Dagar et al. 1991; Debnath 2004; Sreelakshmi et al. 2020a).

Morphological features of *Bruguiera sexangula*. (A) Habit; (B) Bark; (C) stem base with buttresses; (D) Leaves; (E) Flower bud; (F) ridged yellowish calyx tube; (G) Petals and stamens; (H) Bilobed petal with short petal pines and apex without petal bristlers; (I) smooth mature propagule with blunt tip.

Bruguiera cylindrica (L.) Blume, Enum. Pl. Javae 1: 91 (1827).

Bruguiera cylindrica is common in Indian mangroves. It is small tree growing in inner mangroves and occasionally forms pure stands that appear similar in appearance to those of *B. parviflora*. *Bruguiera cylindrica* is distinguished from other *Bruguiera* species by its small flowers, multi flowered inflorescences and calyx with fully reflexed calyx lobes.

Mature propagules

KEY CHARACTERS

- Tree grows upto 10m height, buttresses and knee roots present
- Bark grey, smooth, slightly fissured
- Leaves are narrowly elliptical, leaf tip acuminate, cuneate at base, petiole reddish green
- Inflorescences axial, three flowered
- Flowers small, calyx tube smooth, calyx lobes reflexed
- Petal bi-lobed, petal spine longer than petal lobe, petal bristles present.
- Propagules green to purple green, smooth, slightly curved

Dark green small leaves

Three flowered Inflorescences

Mature flower with white petal

Species feature

Persistent calyx with reflexed calyx lobes

Bruguiera cylindrica (L.) Blume

Taxonomy

Tree: columnar or spreading, height to 10–15 m, evergreen (A). *Bark*: grey, fissured (G); *Root*: knee like pneumatophores (J), buttress at base of stem (I). *Leaves*: simple, opposite, broadly elliptic to narrowly obovate (B), broader towards the apex, margin entire, dark green, 10–13 × 4–7 cm, ratio of length to width >2, apex acute to acuminate, base cuneate; petiole reddish mostly, green also, 2.5–3.5 × 0.3 cm. *Inflorescences*: axillary, 3 flowered, peduncle 0.5–1cm long (C). *Mature flower bud*; cylindrical, greenish white, 1–1.2 × 0.4 cm; calyx 8 lobed, calyx tube 0.6 cm long, lobes 0.5 cm long; petals creamy white, 0.5 cm long, bilobed, apex with 2–3 bristles, 0.2 cm long, petal spine exceeds the lobe, 0.4 cm long (D); stamens 16, 2 enclosed in each petal; style 0.3–0.5 cm long, stigma 3 lobed (E). *Mature Fruits*: present within the calyx tube, viviparous germination. *Mature hypocotyls*: green, brown on maturity, slightly curved and grooved (H), 10–15 × 0.5–0.8 cm, tip blunt, calyx persistent, lobes reflexed (F); plumule 0.2 cm long.

Distribution

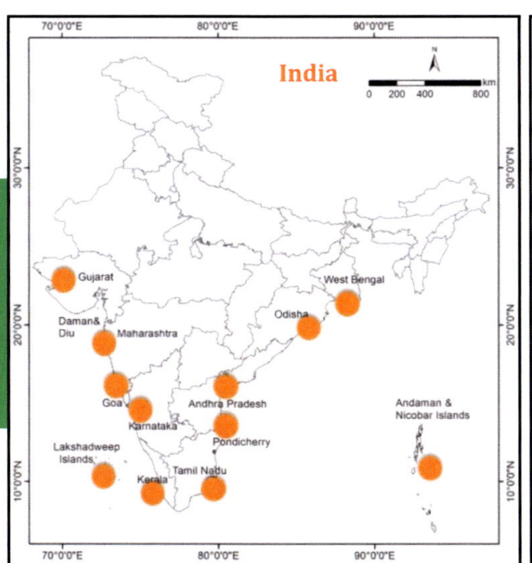

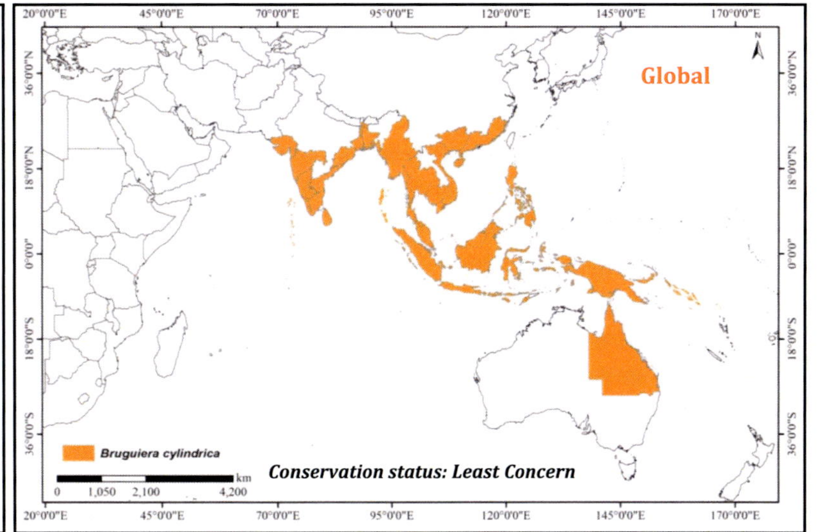

Habitat and Ecology

Often occurs in intermediate and upstream position. It can also occur at landward margin along with *Acrostichum* spp., *Avicennia* spp.

Phenology

Flowering and fruiting occur throughout the year. Flowering peaks from November to February; Fruiting peaks from March to May.

Notes

The size of propagule is highly variable in *B. cylindrica.* Population growing in high saline areas produces small (<10 cm) propagules. It is often abundant in polluted mangroves areas (e.g. Mumbai) in association with *Avicennia marina.*

Morphological features of *Bruguiera cylindrica*. (A) Habit; (B) leafy rosette with propagules; (C) three flowered inflorescences; (D) flower with white petal having long petal spine and bristles; (E) style and stamen; (F) calyx with reflexed lobes; (G) finely fissured bark; (H) mature propagules; (I) buttress; (J) knee roots.

Flower of *Bruguiera parviflora*

Genus: *Ceriops* — Rhizophoraceae

Genus *Ceriops* represents five species globally viz., *C. australis*, *C. decandra*, *C. pseudodecandra*, *C. tagal* and *C. zippeliana*. Two species namely *C. tagal* and *C. decandra* are known from India. *Ceriops* species can be easily distinguished from other Rhizophoraceae members by its blunt calyx lobes, yellowish green leaves which are ovate to slightly obovate or elliptic-oblong, apices rounded or slightly emarginate, never apiculate and a fruiting drupe with viviparous propagule. *C. decandra* is easily distinguished from *C. tagal* by its short peduncle, downwards facing calyx lobe, sharply ridged short hypocotyl warty towards apex and hypocotyls always in upright position. *C. tagal* possesses long peduncle, long propagules with yellow colour collar and calyx lobes facing upwards.

Key to *Ceriops* species

1. Peduncle >1cm, calyx tube flattened, calyx lobes reflexed, petal bristles 3, propagules drooping .... ***Ceriops tagal***
2. Penduncle <1cm, calyx tube cup shaped, calyx lobes not reflexed, petal bristles many, propagules upright.........***Ceriops decandra***

Ceriops tagal

Inflorescences with long peduncle

Flattened calyx with reflexed lobes

Ceriops decandra

Inflorescences with short peduncle

Calyx lobe facing downward

Ceriops tagal (Perr.) C.B.Rob., Philipp. J. Sci., C 3: 306 (1908).

Rhizophoraceae

Ceripos tagal is one widely distributed mangrove species in India. However, its abundance is high in A & N Islands compared to west and East Coasts of mainland India. Generally found in inner mangrove area and often grows as broad monotypic stands. *Ceriops tagal* is easily recognized in field by its erect spathulate leaves, long peduncle, sunken calyx tube, calyx lobes facing upwards and ribbed slender hypocotyls

KEY CHARACTERS

- Tree grows upto 5–10m height, buttress and knee roots present
- Bark grey in colour, smooth
- Leaves are obovate, rounded leaf tip, cuneate at base, yellowish green
- Inflorescences axial, peduncle long (>1cm), multiflowered
- Flowers small, calyx 5 lobed, green; petals 5, white turns brown on maturity, apex with 3 petal bristles
- Propagules green to brown, warty throughout, ridged and grooved, terminally reddish purple, always drooping, calyx persistent flattened with reflexed lobes

Inflorescences with long peduncle

Flower with 5 lobed calyx with tiny petals

Bark orange- pink with raised lenticels

Species feature

Mature propagules having grooves and ribbing

Ceriops tagal (Perr.) C.B.Rob.

Taxonomy

Tree: spreading, medium sized height to 10 m (A). *Bark*: orange brown with dark lenticels (F). *Root*: knee like pneumatophores (J), buttresses at stem base. *Leaves*: simple, opposite, yellowish green on sunlight, dark green under shade, obovate, base cuneate, apex rounded (B, C), 4–10 × 2–6 cm, margin entire; petioles yellowish green to green, 1.5–3 × 0.1–0.3 cm. *Inflorescences*: axillary, 2–12 flowered, peduncle 1–2 × 0.2–0.3 cm (G); *Mature flower buds*: oblong, 0.4–0.6 × 0.3 cm (D); calyx lobes 5, 0.4–0.5 × 0.2–0.3 cm; petals 5, creamy white when young, brown on maturity, 0.3 × 0.1 cm, apex emarginated with three clavate appendages (H); stamens 10, each petal enclosed two stamens, 0.3 cm long (I); style slender, seated on domed ovary, 0.2–0.3 cm long (E). *Mature Fruits*: inverted pear shaped, brown, 2–3 × 1 cm, persistent with calyx, calyx lobes reflexed (K, L). *Mature hypocotyls*: pencil like but tapered, ridged, green, tip pointed, distal half widest, 10–20 × 0.5–1 cm (M); plumule 0.5 cm long.

Distribution

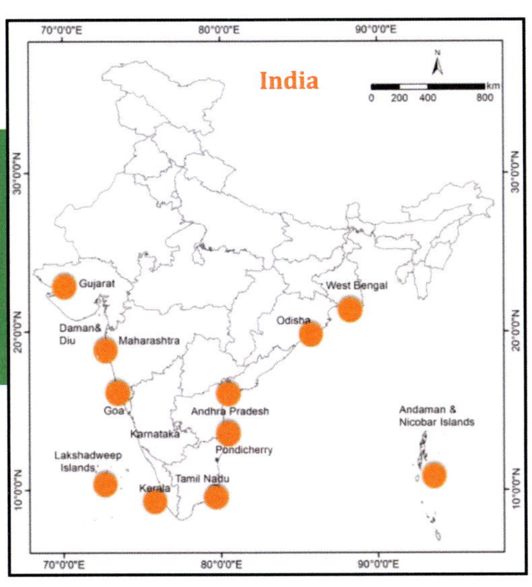

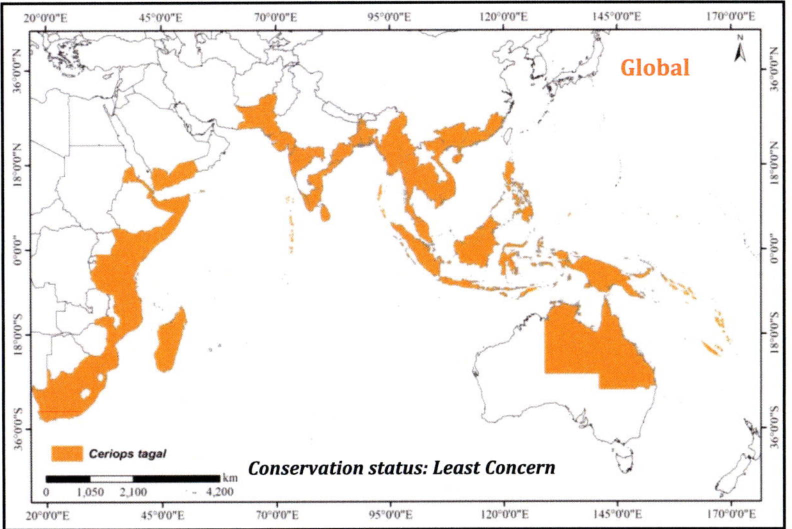

Habitat and Ecology

Often found in intermediate and downstream estuarine position in association with *Bruguiera* spp., *Rhizophora* spp., and *Xylocarpus* spp. Under shade leaves become dark green whereas leaves become yellowish green on exposure to sunlight.

Phenology

Flowering and fruiting occur throughout the year. Flowering peaks from May to June; Fruiting from August to November.

Notes

The absence of *C. tagal* and presence of *C. decandra* along the coast of Karnataka as reported earlier, need to be verified.

Morphological features of *Ceriops tagal*. (A) Habit; (B) leafy rosette with flowers; (C) obovate leaves; (D) mature bud; (E) style; (F) bark; (G) inflorescences with long peduncle; (H) white petal with three bristles; (I) stamens; (J) knee roots; (K &L) fruits with reflexed calyx lobes; (M) propagule with ridges.

Ceriops decandra (Griff.) W.Theob., Burmah ed. 3, 2: 480 (1860). Rhizophoraceae

Ceripos decandra is most common mangrove species along the East Coast of India. Its occurrence along the west coast of India remains doubtful. *Ceriops decandra* is easily distinguished from *C. tagal* by its inflorescences with short peduncle, cup shaped calyx, petals with numerous bristles and upright propagules. It is difficult to distinguish *C. decandra* and *C. tagal* without flowers and fruits due the wide variation in leaf shape and colour.

Mature propagule

KEY CHARACTERS

- Tree grows upto 2–6m height, buttress and knee roots present
- Bark grey in colour, smooth
- Leaves are oblong obovate, rounded leaf tip, cuneate at base, yellowish green
- Inflorescences axial, peduncle long (<1cm), multiflowered
- Flowers small, calyx 5 lobed, green; petals 5, white turns brown on maturity, apex with numerous petal bristles
- Propagules green to brown, warty throughout, ridged and grooved, terminally reddish purple, always upright, calyx persistent cup shaped, lobes facing downward

Immature propagule

Inflorescences with short peduncle

Petal with numerous bristles

Cup shaped calyx

Ceriops decandra (Griff.) W.Theob.

Taxonomy

Tree or shrub: spreading, 2–6m high (A). *Bark*: light-grey, peeling off into thin flakes. *Root*: knee like pneumatophores, buttresses at stem base. *Leaves*: simple, opposite, yellowish green on sunlight, dark green under shade, oval to obovate (B), base cuneate, apex rounded, 4–9 × 2.5–6 cm, margin entire; petioles yellowish green to green, 1.2–1.8 × 0.1–0.3 cm. *Inflorescences*: axillary, 4–16 flowered (C), peduncle 0.5-1 × 0.2–0.3 cm; *Mature flower buds*: oblong, 0.3–0.5 × 0.3 cm; calyx lobes 5, 0.3–0.5 × 0.2–0.3 cm; petals 5, creamy white when young, brown on maturity, 0.3 × 0.1 cm, apex emarginated with numerous clavate appendages (D); stamens 10, each petal enclosed two stamens, filament 1.6–2.0 mm long; anther 1–1.2 mm long, ovoid, dorsifixed, with one long connective protrusion; style slender, seated on domed ovary, 0.2–0.3 cm long; ovary inferior, 3-locular, 2 ovules in each locule. *Mature Fruits*: ovoid, 0.6–1.0 × 0.5–0.6 cm, persistent with cub shaped calyx (E-G), calyx lobes facing downward (G). *Mature hypocotyls*: pencil like but tapered, often in upright position (H, I), ridged, green, tip blunt, distal half widest, 8–13 × 0.5–0.7 cm.

Distribution

India

Gujarat, Daman& Diu, Maharashtra, Goa, Karnataka, Kerala, Lakshadweep Islands, Tamil Nadu, Pondicherry, Andhra Pradesh, Odisha, West Bengal, Andaman & Nicobar Islands

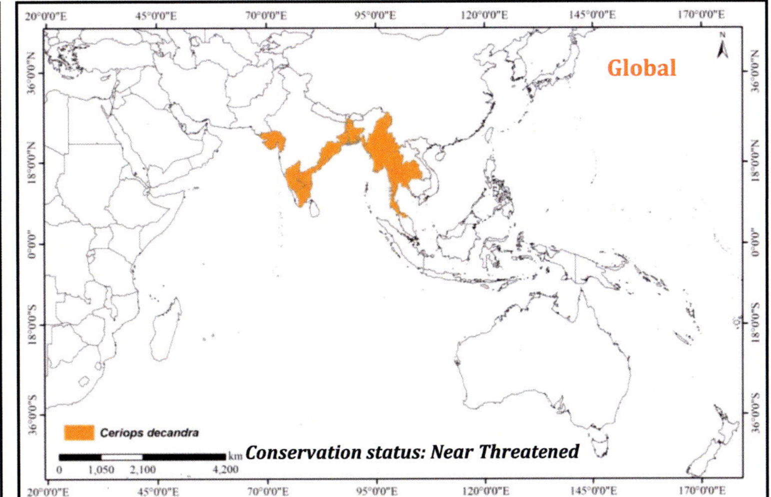

Habitat and Ecology

Often found in inner mangroves as understory at mid and upstream estuarine position or mid and low intertidal region.

Phenology

Flowering and fruiting occur throughout the year.

Notes

The reports of *C. decandra* from A & N Islands are erroneous (e.g. Sreelakshmi et al. 2020a). Two close relatives of *C. decandra* viz., *C. zippeliana* and *C. pseudodecandra* are distinguished based on pubescence of petal margin. Petals margin of *C. decandra* is entire and hairy, whereas partially hairy in *C. pseudodecandra* and without hairs in *C. zippeliana* (Sheue et al. 2010).

Morphological features of *Ceriops decandra*. (A) Habit; (B) Leaves; (C) Inflorescences; (D) petals with numerous bristles; (E) short peduncle; (F) Cup shaped calyx; (G) Infructescence with short peduncle; (H) upright propagules; (I) propagule with ridges.

Bruguiera gymnorhiza

Genus: *Cynometra*

Fabaceae

The genus *Cynometra* L. belongs to the family Fabaceae and has about 85 species in the tropical regions (Mabberley 2008). The genus is characterized by uni-jugate or bi-jugate leaves, small bud scales, 4 calyx segments, 5 petals, 10 stamens and 1 seeded fruit. In India, *Cynometra* is represented by seven species viz., *C. beddomei*, *C. bourdillonii*, *C. iripa*, *C. malaccensis*, *C. ramiflora*, *C. travancorica* and *C. cauliflora* (Sanjappa 1992). Of these, *C. iripa* and *C. ramiflora* are often classified as true mangrove species in India (Naskar 2004; Kathiresan 2008; Sanjappa et al. 2011), but only *C. iripa* is the most common representative in mangroves swamps (Meeuwen 1970) and classified as a true mangrove species globally (Duke 1992; Polidoro et al. 2010) whereas *C. ramiflora* is found commonly in the fringes of the fresh water stream or back mangroves and also found inland. *Cynometra iripa* is known from Sundarbans, Odisha, Maharashtra and A & N Islands, whereas *C. ramiflora* known only from A & N Islands (Ragavan et al. 2017).

Key to *Cynometra* species

1. Fruits with prominent lateral beak, flowers with bent style, densely hairy ovary and leaves with off centred mid vein and obtuse slightly emarginate tip.......................... ***Cynometra iripa***
2. Fruits with sub terminal beak, flowers with straight style, glabrous ovary and leaves with off centred mid vein and pointed slightly emarginate tip................................ ***Cynometra ramiflora***

Cynometra iripa

Cynometra ramiflora

Cynometra iripa **Kostel.**, Allg. Med.-Pharm. Fl. 4: 1341 (1835).

Fabaceae

Cynometra iripa is distinguished from *C. ramiflora* by its eccentric bent style, hairy ovary, distally curved calyx lobes and fruit with prominent lateral beak whereas in *C. ramiflora* style is straight, ovary curly hairy outside and glabrous inside, calyx lobes erect and fruits with sub terminal beak (Meeuwen 1970; Giesen et al. 2006; Ragavan et al. 2017). It is known to occur in Sundarbans, Odisha, Maharashtra and Andaman Islands.

Bent style with hairy ovary

KEY CHARACTERS

- Medium sized tree, no above ground roots
- Bark smooth, mottled greenish brown with numerous lenticels.
- Compound leaves mostly have 2 pairs of leaflets, middle vein is slightly off-center
- Inflorescence axial, capitate, multiflowered
- Flowers are small and white, style bent prominently, and ovary is hairy
- Fruits are hard, almost globular pods with a prominent lateral beak. Surface is deeply wrinkled and covered with short hairs

Inflorescences

Species feature

Leaves with off centered mid vein

Wrinkled pod with lateral beak

Flower with recurved sepals

Cynometra iripa Kostel.

Taxonomy

Tree: spreading, medium sized height 3–10 m, evergreen (A). *Bark*: smooth, brown-grey patchy finely fissured (C). *Roots*: not above ground, stem base simple (I). *Leaves:* compound, alternate, 2 pairs of leaflets (D); leaflets are green, mid vein off centred, elliptic, apex obtuse with notched tip, base cuneate (B), 2–6.5 × 1–3 cm, lower pair often smaller (D); leaflet stalk minute, 0.2 cm long. *Inflorescence*: axillary, capitate, 3–7 flowered (E); *Mature flower buds*: globular, bisexual, covered by bract; peduncle 1cm long; calyx 4 lobed, 0.5 cm long, reflexed (H); petals 5 creamy white, lanceolate, 0.5–0.8 cm long, reflexed (F); stamens 10, filament slender 0.5–0.7 cm long, anthers small (F); ovary hairy outside and inside (G), inserted eccentrically on a short stalk; style slender, glabrous, bent (G), up to 0.5 cm long, stigma capitate. *Mature fruit*: one seeded, woody, wrinkled with distinct lateral beak (I), green turn to brown on maturity, 2–3.5 × 1.5–2.5 cm.

Distribution

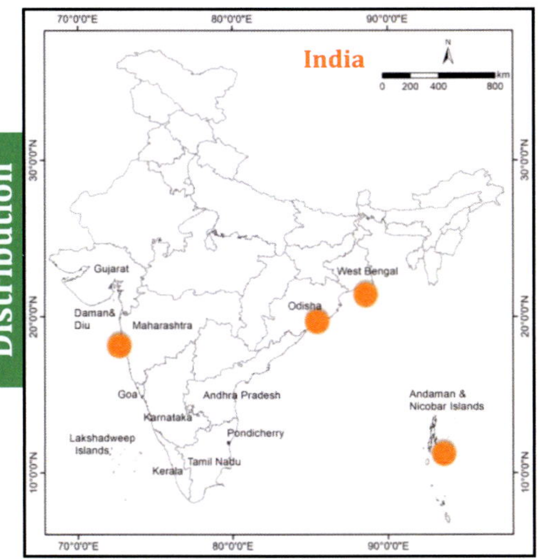

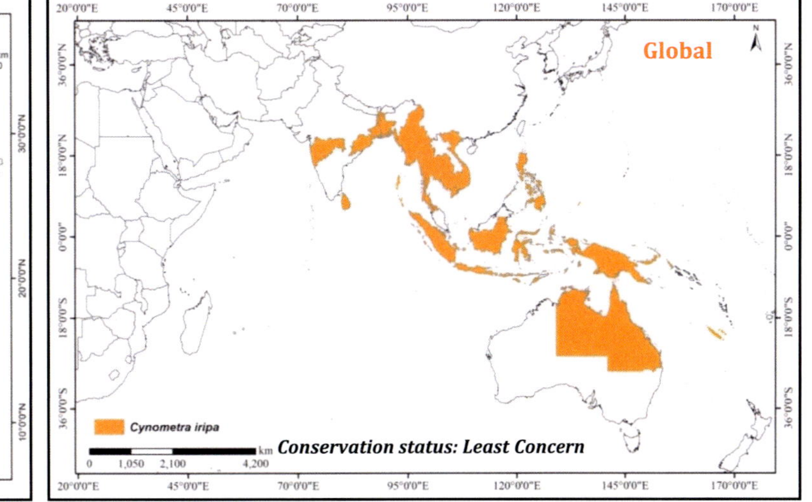

Habitat and Ecology

Often occur in landward margin of mangrove in association with *Pheonix paludosa*, *Heritiera littoralis*

Phenology

Flowering August to December; Fruiting January to April.

Notes

Earlier *C. iripa* was reported as a variety of *C. ramiflora* viz., "var. mimosoides" and "subsp. *bijuga*. The nomenclature ambiguity, identity and distribution of *C. iripa* and *C. ramiflora* are resolved (Ragavan et al. 2017).

Morphological features of *Cynometra iripa*. (A) Habit; (B) leaflets with blunt emarginate tip; (C) bark; (D) leaf with two pairs of leaflets; (E) inflorescences; (F) flowers; (G) hairy ovary and bent style; (H) reflexed calyx lobes; (I) stem base; (J) wrinkled pod with lateral beak.

Cynometra ramiflora L., Sp. Pl. 1: 382 (1753).

Cynometra ramiflora is highly variable. This species and its synonyms based on differences in the number and size of leaflets often cause confusion with *C. iripa*. *Cynometra ramiflora* is distinguished from *Cynometra iripa* by its straight style, curly hairy ovary outside and glabrous inside, erect calyx lobes and fruits with sub terminal beak, while *C. iripa* has eccentric bent style, hairy ovary, distally curved calyx lobes and fruit with prominent lateral (Meeuwen 1970; Giesen et al. 2006). It has restricted distribution only in Andaman Islands and not present in mainland India.

Flower with erect petal

KEY CHARACTERS

- Medium sized tree, no above ground roots
- Leaves are compound, 2 pairs of leaflets, lower pair often smaller, terminal leaflets much larger, mid vein off-center
- Inflorescence axillary, capitate, multiflowered
- Flowers are small white, style straight and has pubescence, ovary is curly hairy
- Fruit are pod like with sub-terminal beak, green turn to brown on maturity, one seeded, woody, wrinkled.

fruit with sub-terminal beak

Leaf with two pairs of leaflets

Pointed emarginate tip

Inflorescences

Cynometra ramiflora L.

Taxonomy

Tree: spreading, medium sized height 3–10 m, evergreen (A). *Bark*: smooth, brown-grey and patchy with numerous lenticels (E). *Roots*: not above ground, stem base simple. *Leaves*: compound, alternate, 2 pairs of leaflets; leaflets are green, mid vein off centered, elliptic, apex acute with notched leaf tip (C), base cuneate, 3.5–15 × 2–5.5 cm, lower pair often smaller, terminal leaflets much larger (B, D); leaflet stalk minute, 0.2 cm long. *Inflorescence*: axillary, capitate, 3–10 flowered (G); *Mature flower buds*: globular, bisexual covered by bract; peduncle 1cm long; calyx 4 lobed, white, 0.5 cm long, reflexed (F); petals 5 creamy white, lanceolate, up to 0.6–1 cm long, not reflexed (F) as in *C. iripa*; stamens 10, filament slender 0.7–1 cm long, anthers small; ovary curly hairy outside, glabrous inside, inserted eccentrically on a short stalk; style slender, straight, hairy at base (F), 0.5–0.8 cm long, stigma capitate. *Mature fruit*: one seeded (K), woody, wrinkled with sub-terminal beak (I), green turn to brown on maturity (H, J), 2–2.5 × 1–1.5 cm.

Distribution

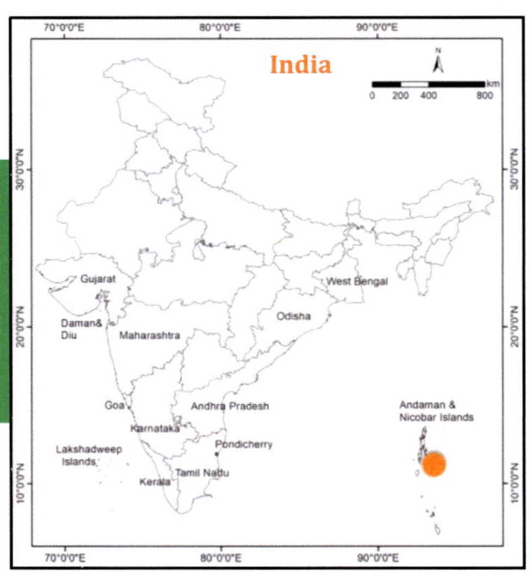

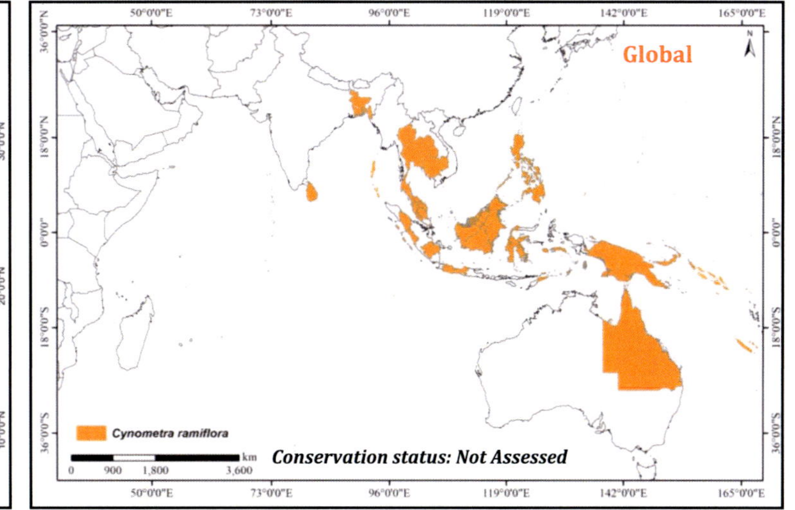

Habitat and Ecology

Found on landward margins of mangroves, on the banks of the freshwater stream

Phenology

Flowering December to January; Fruiting January to February.

Notes

Reports of *C. ramiflora* from mainland India as mangrove species is erroneous. *Cynometra ramiflora* sub sp. *genuina* and subsp. *heterophylla* are synonyms of *C. ramiflora*.

Morphological features of *Cynometra ramiflora*. (A) Habit; (B) leaf with two pairs of leaflets; (C&D) leaflets with pointed emarginate tip; (E) bark; (F) flower; (G) inflorescences; (H) fruit cluster; (I) fruit with sub-terminal beak; (J) mature fruit; (K) seed.

Dolichandrone spathacea Seem., J. Bot. 1: 226 (1863).

Bignoniaceae

Dolichandrone spathacea is the only member of the family Bignoniaceae occurring in mangrove habitat. It is a frequent constituent of the high intertidal mangroves. It is easily recognized in the field by its opposite, pinnately- compound leaves and characteristically large, conspicuous flower with a tubular, trumpet shaped corolla. *Dolichandrone spathacea* is abundant in Andaman Islands whereas it has restricted distribution in East and West Coasts of mainland India..

KEY CHARACTERS

- Tree grows up to 20m height
- Bark grey to dark brown, fissured in matured tree
- Leaves opposite, compound, shiny dark green above than below.
- Leaflets with 2–4 pairs of ovate -lanceolate leaf with acuminate apex
- Inflorescence terminal racemes
- Flowers large, conspicuous with tubular, trumpet shaped white corolla
- Fruits are long horn like with hard, leathery partitions having numerous dark grey seeds arranged in many rows

Long bean Matured fruits

Species feature

Compound leaf with floppy wrinkled leaflets

Smooth, grey colored young bark

Long bean like ripened fruits

Trumpet shaped white flower with long corolla tube

Taxonomy

Tree: spreading, height to 15 m, evergreen, irregular in shape, trunk short (A). *Bark*: grey to brown, fissured on maturity (J). *Roots*: not above ground, stem base simple. *Leaves*: compound, opposite, 8–30 cm long; petiole 3–10 cm long; leaflets 2–4 pairs, wrinkled, ovate oblong to lanceolate (B), mid vein is off-centered, tip pointed, base irregular, 5–11 × 2–6 cm, margin entire, minutely hairy; leaflet stalk short, 0.5-1.5 cm long. *Inflorescence*: terminal, 2-8 flowered; peduncle 1.5–2 × 0.5 cm. *Mature flower buds*: flowers are conspicuous, 10–15 × 0.5–1 cm; bracteole green, 5–8 cm long, apex with blunt mucronate (D); calyx minute; corolla trumpet shaped, white (C), 5 lobed, 2.5–3 cm long, corolla tube greenish white, 10 × 0.5–1 cm; stamens 4, attached at the base of corolla lobe (E, F), 2–2.5 cm long, fifth one is vestige; style long up to 15 cm long, exceeding beyond the stamens (H). *Mature Fruits*: bean like, green turns dark brown on maturity (G), flattened 30–60 × 2–3 cm; seeds rectangular, winged, and arranged in many rows, 0.5–1 × 1–2 cm.

Dolichandrone spathacea Seem.

Morphological features of *Dolichandrone spathacea*. (A) Habit; (B) leaf; (C) flower; (D) bracteole; (E & F) stamens; (G) fruits; (H) style; (J) small fruits; (J) finely fissured bark.

Distribution

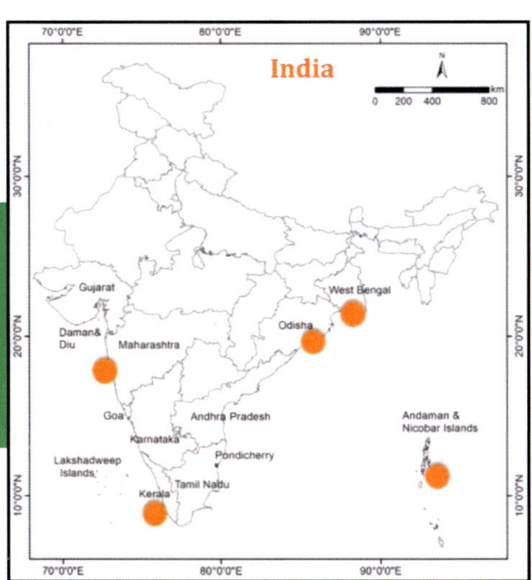

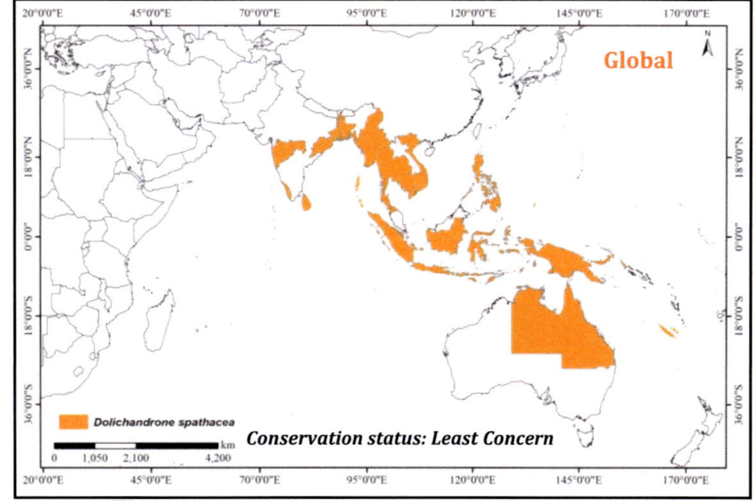

Habitat and Ecology

Often found in landward margin of mangroves, in upstream estuarine position it occurs in association with *Excoecaria agallocha, Heritiera littoralis.*

Phenology

Flowering and fruiting occur throughout the year, but it varies depending on maturity of the tree. Flowering peaks from March to May and fruiting from June to July.

Notes

Dolichandrone spathacea is often categorised as mangrove associate. But this species forms pure stands in association with *Heritiera littoralis* and *Excoearia agallocha* in undisturbed mangrove areas This species produces cable roots in the habitats with regular tidal inundation..

Stilt roots of *Rhizophora apiculata*

Genus: *Excoecaria*

Genus *Excoecaria* L. belongs to the family Euphorbiaceae Juss. and is commonly recognized with mangrove representatives (Duke 2006). *Excoecaria* is distinguished from its closely related members by a combination of characters, including the deciduous condition, axillary inflorescences, male flowers with 3 stamens, and the absence of caruncle from the seed (Tomlinson 1986). The genus has up to 40 species in the IWP regions from tropical Africa and Asia to western pacific. Two occur in mangroves including *Excoecaria indica* (Wild.) Muell. -Arg and *Excoecaria agallocha* L. In India *E. agallocha* is abundant, known to occur in both east and west coast, whereas *E. indica* is rare, known to occur in Sundarbans, Odisha, Kerala and A & N Islands (Ragavan et al. 2015a).

Leaf of *E. agallocha* with wavy margin

Key to *Excoecaria* species

1. Leaf margin entire to wavy, bark fine fissured with lenticels, fruits three lobed ....***Excoecaria agallocha***
2. Leaf margin serrated, bark smooth, trunk with thrones, fruits rounded with three seeds.........................***Excoecaria indica***

Crenulate-lanceolate leaves of *E. indica*

Female flower of *E. agallocha* with three style

Male flowers of *E. agallocha* with three stamens

White milky latex of *E. indica*

Excoecaria agallocha L., Syst. Nat., ed. 10. 2: 1288 (1759).

Excoecaria agallocha is abundant in Indian mangroves and it is conspicuously dioecious with separate male or female tree. Leaves shape and size of *E. agallocha* exhibit wide variation among the population. During dry season leaves of *E. agallocha* turn bright red or orange before they fall. Leaves of *E. agallocha* have wavy margin whereas in *E. indica* leaf margin is serrated. In *E. agallocha* male inflorescence of male tree is relatively longer than female inflorescence of female tree.

Three lobed small fruits

KEY CHARACTERS

- Tree grows up to 10m height, dioecious
- No above ground root; but can sometimes have spreading surface roots
- Bark grey and rough
- Leaves light green to dark green turn reddish before senescence, leaves exude white sap when broken
- Inflorescences catkin like, male inflorescences longer than female
- Fruits 3-lobed and contain dark brown seeds

Species feature

Leaves

Matured male inflorescences

Immature male inflorescences

Female inflorescences

Above ground roots

Excoecaria agallocha L

Taxonomy

Tree: spreading, dioecious, deciduous, height 10–15 m (A). *Bark*: grey, rough, fissured on maturity, pustular with lenticels (K). *Root*: above ground, serpentine like (A), stem base simple, occasionally slight buttresses. *Leaves*: simple, alternate, elliptic (B), green, mature leaves red in colour (C) 5–15 × 2–6 cm, ratio of length to width >2, apex acute to acuminate, base cuneate, margin variably serrate to entire, often wavy (D); petiole green, terete 1–3.5 × 0.2 cm. *Inflorescences*: axillary, unisexual; male inflorescence 5–10 cm long (E), flowers arranged spirally with bract (H), calyx lobes 3; stamens 3, yellow, 0.3-0.5 cm long (G), pistillode absent; female inflorescence 2–4 cm long (F), peduncle 0.5 cm long, bract glandular, basal bracteoles 2, calyx lobes 3, staminodes absent, ovary tri-locular; style 3, 0.3 cm long (I). *Mature fruit*: 3-lobed capsule (J), 1–1.5 cm wide, green becoming brown on maturity, three seeded; seeds are rounded, black or dark brown, streaked, up to 0.3–0.5 cm in diameter.

Distribution

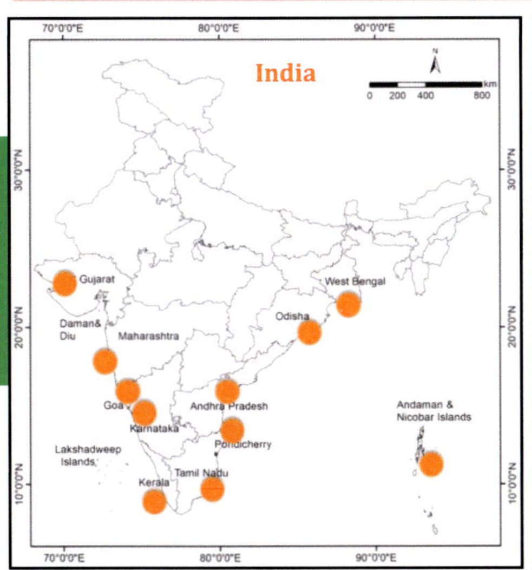

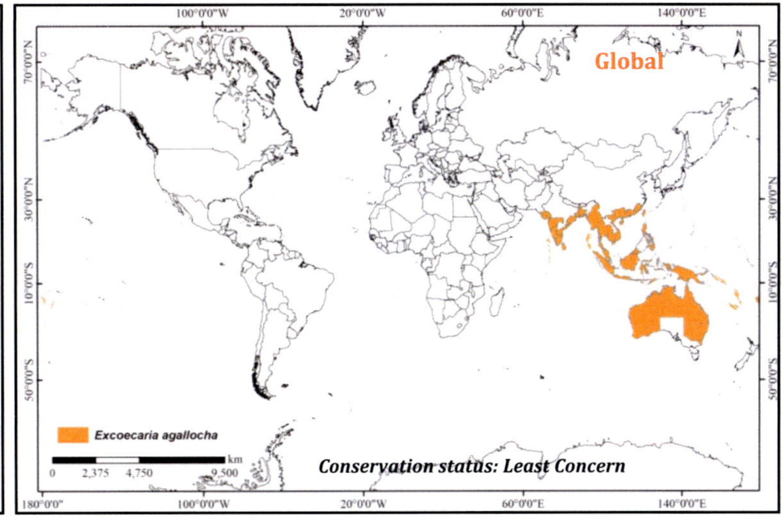

Habitat and Ecology

Often found on the landward margin of mangroves, it can also occur in intermediate and upstream estuarine position in association with *Heritiera littoralis, Avicennia marina* and *Sonneratia apetala*.

Phenology

Flowering June to August; Fruiting September to November.

Notes

Excoecaria agallocha is commonly called as "Milky Mangrove" due to latex secretion. This species is worshipped around the world especially as the temple tree in Chidambaram of Tamil Nadu, for the last 17 centuries (Kathiresan 2018). The plant produces a toxic latex, that is capable of blinding the human eyes, and hence it is also called as 'blinding tree'.

Morphological features of *Excoecaria agallocha*. (A) Habit; (B) leafy rosette; (C) mature leaf; (D) leaf with wavy margin; (E) male inflorescences; (F) female inflorescences; (G) male flowers with three stamens; (H) young male inflorescences; (I) female flowers with three style; (J) three lobed fruits; (K) bark.

Excoecaria indica Müll.Arg., Linnaea 32(1): 123 (1863).

Euphorbiaceae

Excoecaria indica is rare in India and known to occur in Sundarbans, Odisha, Kerala and A & N Islands. Moreover, due to lack of inadequate species-specific information, *E. indica* is at the IUCN category of data deficient and it is critically endangered in India. *E. indica* is distinguished from *E. agallocha* by its thorny trunk, crenulate-lanceolate leaves and green fruits the size of cherry with three seeds. The distribution and habitat ecology of *E. indica* is poorly known.

Terminal Inflorescences

Key characters

- Tree grows up to 10-15m height with upright branches and more or less drooping twigs
- Thorny trunk without buttress
- Leaves lanceolate, margins conspicuously serrate, with two small glands at the base of the blade
- Inflorescences solitary, raceme-like
- Male or staminate flowers are numerous with 3 prominent stamens
- Female or pistillate flowers are 1 or 2; solitary with 3 long styles
- Fruit is a round, woody capsule, green; black when dries, 3-seeded.

Species feature

Thorny trunk

Smooth rounded fruits

Cross section of fruits showing three locus arrangement

Matured fruits

Matured Seeds

Excoecaria indica Müll.Arg.

Taxonomy

Tree: spreading, multi-stemmed, monoecious, deciduous, branches upright with drooping twigs, height 10–15 m (A). *Bark*: fissured, greyish brown in colour, trunk thorny (C). *Roots*: no above ground. *Leaves*: simple, alternate, elliptic or lanceolate (B), 7–14 × 3–4 cm, base obtuse, margins conspicuously serrate (D), apex sub acuminate to acuminate with two small glands at the base of the blade, lateral veins 18–24; petiole 0.7–2 cm long and reddish (L). *Inflorescences*: solitary, bisexual, raceme-like, to 5–10 cm (E), petals absent. *Mature Flowers*: male or staminate flowers are numerous with 3 prominent stamens (F); stamen filaments 0.5-0.6 mm long at anthesis, nearly absent in bud; anthers 0.4–0.5 cm; female or pistillate flowers are 1 or 2; solitary with 3 long styles (G, H); style 1.5-2 cm long, stigmas 0.4–0.6 cm. *Mature fruit*: round, woody capsule, 2.5–3 cm in diameter, green (I), black when dries (J), 3-seeded (K); fruit stalk 0.8–2.2 cm long; seeds 11–13 × 0.7–0.8 cm, keeled on the back, medium to pale brown, not spotted, without caruncle.

Distribution

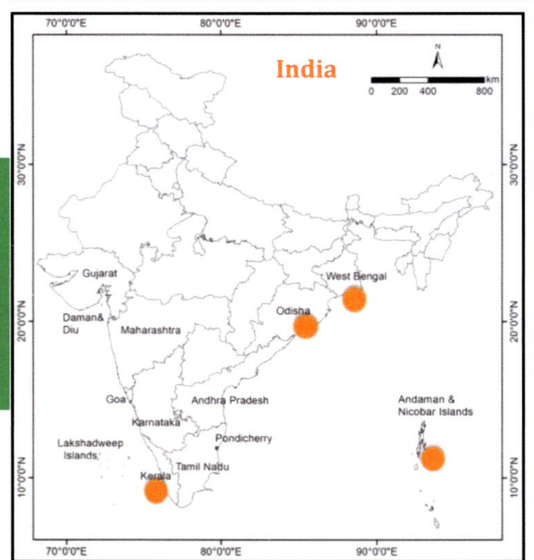

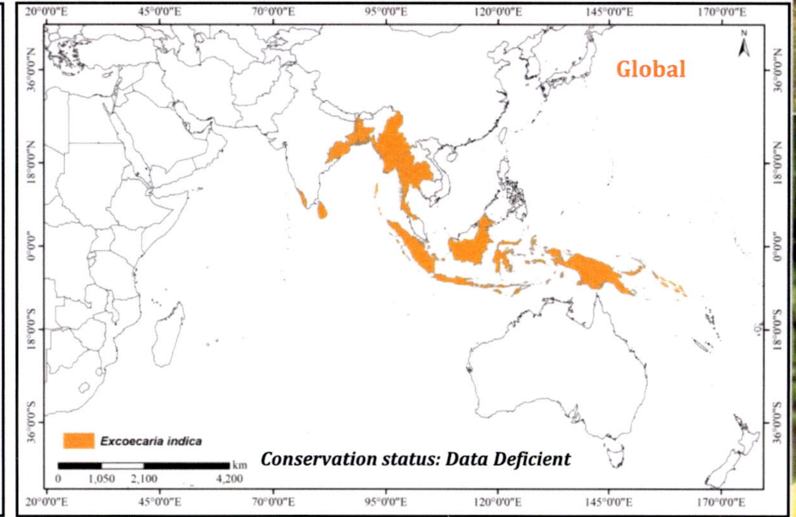

Habitat and Ecology

The habitat ecology of *E. indica* is poorly known. Often occur on the landward margin in association with *Heritiera* spp, *Excoecaria agallocha*, *Dolichandrone spathacea.*

Phenology

Flowering March to June; Fruiting July to October.

Notes

Excoecaria indica is a "Data Deficient" mangrove species. *Excoecaria indica* is monoecious whereas *E. agallocha* is dioecious. *Shirakiopsis indica* (Willd.) Esser is the accepted name of this taxon.

Morphological features of *Excoecaria indica*. (A) Habit; (B) leafy rosette; (C) thorny trunk; (D) lanceolate leaf with serrated margin; (E) inflorescences; (F) male flowers with three stamens (G) female flower with long style; (H) male and female flowers; (I) rounded fruits; (J) ripened fruit; (K) seed; (L) reddish petiole.

Branches of *Heritiera littoralis*

Genus: *Heritiera*

Malvaceae

Two species of genus *Heritiera viz., Heritiera fomes* and *H. littoralis* are known from Indian mangroves, of which the former is reported only from Sundarbans and Odisha, but due to the reduction in freshwater input in both the places it has become rare. *Heritiera littoralis* is known from Odisha, Maharashtra and A & N Islands. In Odisha *H. littoralis* is reported only in Bhitarkanika Wildlife Sanctuary and shows very restricted distribution (Panda et al. 2013). In Maharashtra it is reported by Shaikh et al. (2011), with just nineteen individuals recorded. In A & N Islands it is common in both the groups of Islands (Ragavan et al. 2015). *Heritiera* species are easily recognised in the field by its larger size and silvery grey underside of the leaves. *Heritiera globosa*, which was considered a mangrove species in the past, is excluded from mangroves, as it has been found only at the upper reaches of rivers in Borneo (Duke 2017; Saenger et al. 2019), where the water is always fresh,

Key to *Heritiera* species

1. Above ground roots planks like, fruits with prominent keel on one side..............***Heritiera littoralis***
2. Above ground roots peg-like pneumatophores, fruits with multiple keel................***Heritiera fomes***

Heritiera littoralis Aiton, Hort. Kew. [W. Aiton] 3: 546 (1789). Malvaceae

Heritiera littoralis is most common in A & N Islands, whereas it is restricted to Odisha and Maharashtra in mainland India. It is easily distinguished in the field by its larger seized leaves with silvery white underside and short petiole, smooth ovoid seed pods with an extended keel and stem base with large plank buttresses. Generally found in landward margin of the mangroves i.e., upper estuarine position.

KEY CHARACTERS

- Tree grows upto 10–15m height
- Root buttresses at base of the trunk and plank or ribbon type root
- Leaves are large (up to 30 cm long) dark green leaves that have a silvery white undersurface
- Inflorescences sub-terminal, multiflowered
- Flowers are small and reddish in color
- Fruits are about 7 cm long and have a prominent ridge or keel on one side.

Inflorescences

Leaves

Bunch of flowers at leaf axial

Small reddish color flowers

Fruit with prominent keel

Heritiera littoralis Aiton

Taxonomy

Tree: spreading, monoecious, evergreen, height to 20 m (A). *Bark*: pale grey, fissured (J). *Roots*: above ground, planks like, stem base with prominent buttresses (F). *Leaves*: simple, alternate, oblong elliptic, green above (B) and silvery grey below (C) with pubescent, margin entire, apex acute to obtuse, base rounded, 7–27 × 3–11 cm, ratio of length to width is >2; petiole pale green, bi-pulvinate, 0.5–2.5 × 0.3–0.5 cm. *Inflorescence*: sub-terminal, multi-flowered panicle (E, G), 5-10 cm long, unisexual flowers. *Mature flowers*: 0.3–0.5 × 0.2–0.4 cm; calyx cup shaped with 4–5 short pointed lobes, red hairy inside and pale green hairy outside (I); petals absent; stamens 4–5, fused in male flowers; staminodes minute in female flowers, style 4–5, united, stigma recurved. *Mature fruits*: keeled pod, yellowish green in colour, turns to brown on maturity (D, H), 10 × 5 cm, keel 0.4–0.8 cm long, one seeded.

Distribution

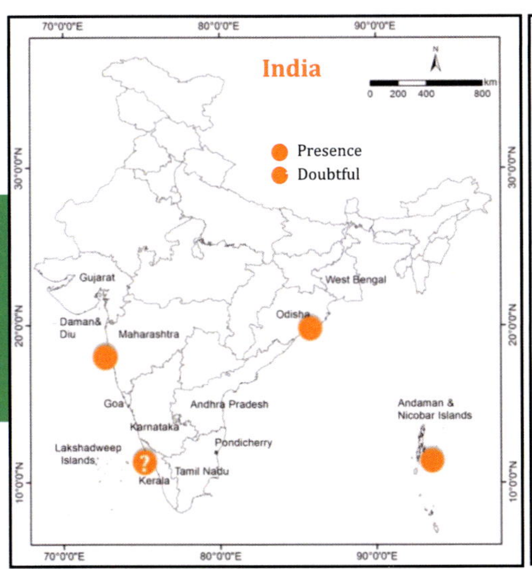

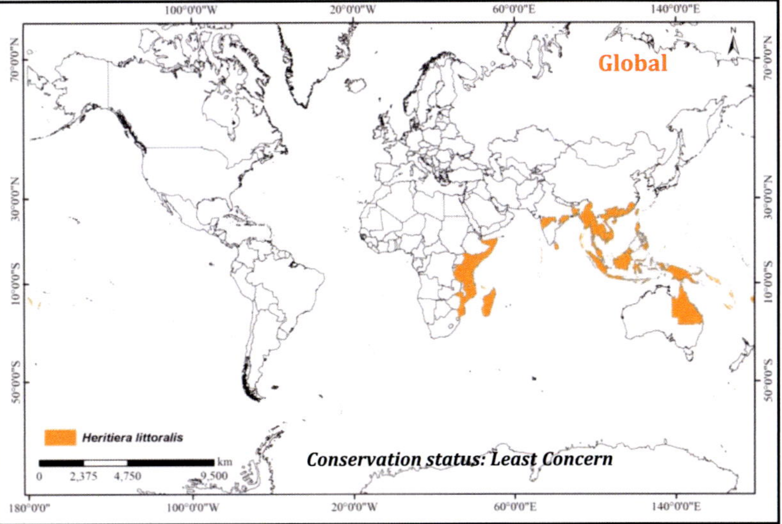

Habitat and Ecology

Often found on the landward margins of mangroves in association with *Excoecaria* spp., *Avicennia* spp., and *Xylocarpus* spp. and may also occupy the fringes of adjacent lowland forest, or rocky shores.

Phenology

Flowering October to December; Fruiting December to March.

Notes

Seedlings of *Heritiera littoralis* can be overlooked as *Brownlowia tersa* in the field due to morphological similarity in leaves viz., silvery grey underside and dark green colour of upper side. Mohandas et al. (2014) noted the occurrence of *H. littoralis* in Kerala coast with taxonomical description, so its occurrences in Kerala need to revaluate.

Morphological features of *Heritiera littoralis*. (A) Habit; (B) leaf; (C) slivery grey underside; (D) keeled fruits; (E) inflorescences; (F) buttress; (G) pink colour flowers; (H) young fruits; (I) calyx; (J) bark.

Heritiera fomes Buch.-Ham, Embassy ed. 2, 3: 319 (1800).

Malvaceae

Heritiera fomes, commonly called as "Sundari". The distribution *H. fomes* is restricted to Sundarbans and Odisha. It is easily identified in the field by its leaves with silvery white underside and fruits with longitudinal ridges. Readily distinguished from *H. littoralis* by the presences of peg-like pneumatophores, whereas *H. littoralis* possess planks root. Generally found in upstream and mid estuarine position in association with *Sonneratia caseolaris*, *Xylocarpus moluccensis* and *Nypa fruticans*.

Bark

KEY CHARACTERS

- Tree grows upto 10–15m height
- Root buttresses at base of the trunk and prominent peg like pneumatophores
- Bark brown to bark brown, smooth when young, turns deeply fissured on maturity
- Leaves elliptic, dark green on upper surface and silvery white on undersurface
- Inflorescences sub-terminal, multiflowered
- Flowers are small and reddish in color
- Fruits are about 5–7 cm long and have a prominent longitudinal ridge

Peg like pneumatophores

Leaves

Leaves

Silver grey underside

Inflorescences

Hairy calyx tube

Heritiera fomes Buch. Ham.

Taxonomy

Tree: spreading, monoecious, evergreen, height to 20 m (A). *Bark*: brown to dark brown, deeply fissured (B). *Roots*: above ground, peg like pneumatophores (C), stem base with prominent buttresses. *Leaves*: simple, alternate, elliptic, dark green above (D) and silvery grey below (E) with pubescent, margin entire, apex acute to obtuse, base cuneate, 5–20 × 3–8 cm, ratio of length to width is >2; petiole pale green, bi-pulvinate, 0.5–2.5 × 0.3–0.5 cm. *Inflorescence*: sub-terminal, multi-flowered panicle (F), 6–11 cm long, unisexual flowers. *Mature flowers*: 0.2–0.5 × 0.2–0.3 cm; calyx cup shaped with 4–5 short pointed lobes, red hairy inside and pale green hairy outside (G); petals absent; male flower has 5 stamens that are fused together, female flower has 4–5 carpels that are loosely attached, style is terminal (G), long, white but after maturation it becomes brown. *Mature fruits*: pod like with longitudinal keels or ridges, yellowish green in colour, turns brown on maturity, 7 × 3 cm, one seeded.

Distribution

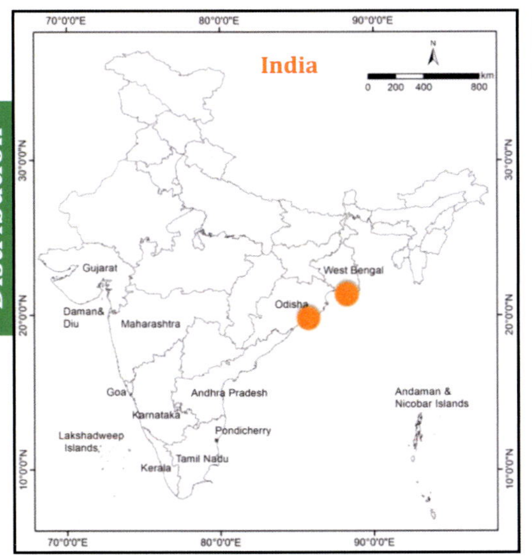

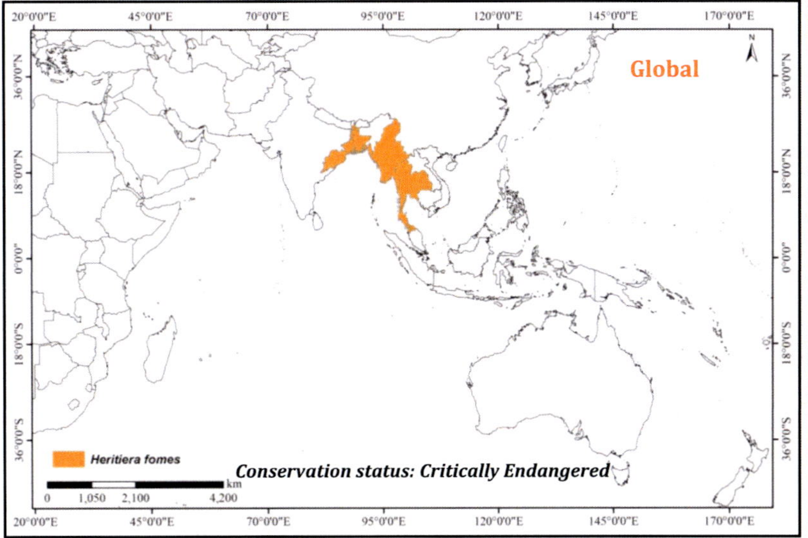

Habitat and Ecology

Often found in mid and upstream estuarine position. It is also present in landward edges of mangroves and along brackish tidal streams

Phenology

Flowering June to July; Fruiting August to September

Morphological features of *Heritiera fomes* (A) Habit; (B) deeply fissured bark (C) peg like pneumatophores (D) dark green leaves (E) silvery grey underside of the leaf (F) Inflorescences; (G) pink colour flower.

Notes

Heritiera fomes is a globally threatened mangrove species. Since presence of peg-like pneumatophores is the key characteristic feature of both *Xylocarpus moluccensis* and *Heritiera fomes,* care should be taken to identify this taxon based on the aerial root. *Heritiera kanikensis* reported by Majumdar and Banerjee (1985) as a new mangrove species from the Bhitarkanika of Odisha is only an ecological variant of *H. fomes* (Kathiresan 2010).

Kandelia candel (L.) Druce, Bot. Exch. Club Soc. Brit. Isles 3: 420 (1914). Rhizophoraceae

Genus *Kandelia* represents two species in global mangroves viz., *K. candel* and *K. obovata*, of which, *K. candel* is common in Indo-Malesia, whereas *K. obovata* is restricted to coast of South China Sea. In India *K. candel* is widely distributed along the east and west coast of Mainland India. The reports of *K. candel* from A & N Islands are erroneous. In the field *K. candel* is easily identified by its elliptic oblong leaves, long peduncle, flower with long and reflexed calyx lobes and propagule with sharp tip.

KEY CHARACTERS

- Medium sized tree grows up to 3–6m, no above ground roots
- Bark rough, brown to dark grey, slightly fissured and flaky
- Leaves dark green, elliptic oblong, apex blunt
- Inflorescences axial, 4-12 flowered, peduncle long, bracteoles V shaped
- Flower buds oblong, calyx lobes reflexed, petals white, deeply bifid with 3–4 long bristles in each lobe, stamens unequal, style does not exceed the stamen
- Propagules 20–40cm long, smooth, tip sharply pointed, calyx persistent, lobes reflexed

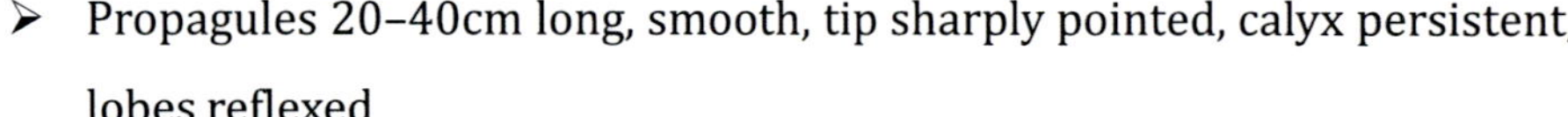

Deeply bifid petal with 3-4 long bristles

Matured propagules

Inflorescences with long peduncle

Flower buds with V shaped bracteoles

Reflexed calyx

Stem base

Bark

Kandelia candel (L.) Druce

Taxonomy

Tree: medium seized, 3-6m high (A). *Bark*: rough, brown to dark grey, slightly fissured and flaky. *Roots*: no above ground roots. Leaves: elliptic oblong, dark green (B), 6–16 × 3–6 cm, obtuse at apex, cuneate at base; petiole 1–1.5 cm long; *Inflorescences*: 2–3 times dichotomously branched, axial (C), 4–12 flowered. *Matured flower bud*: oblong, whitish green, peduncle 1.5–5 cm long; pedicels 0.2–0.3 cm long; bracteoles V-shaped in outline (D), 0.2–0.3 cm long, 2-lobed; calyx 5 lobed, light green on abaxial side and whitish on adaxial side when blooming (E), 1.4–1.6 × 1.9–2.1 cm; petals 5, white, deeply bifid, each lobe with 3–4 long bristles (G); stamens, numerous, unequal (F), filaments 0.8–1.3 cm long, anthers 0.1 cm long; style not exceeds the stamens, 0.9–1.2 cm long, greenish white. *Mature Fruits*:1.5–2.5 cm long, with calyx tube, lobes reflexed. *Mature hypocotyl*: green, smooth, attenuate at apex (H), 20–40 × 7–9 cm.

Distribution

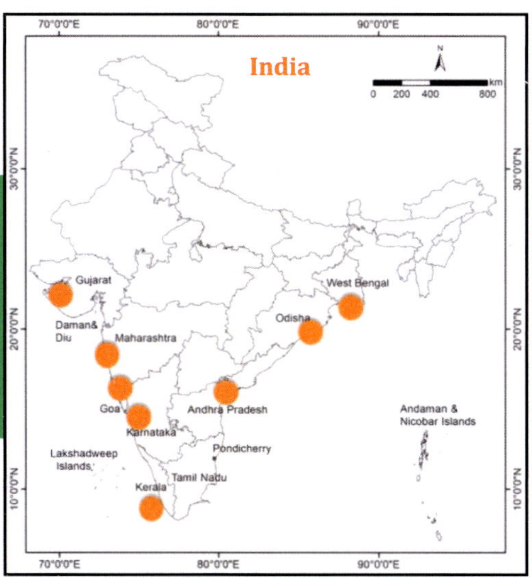

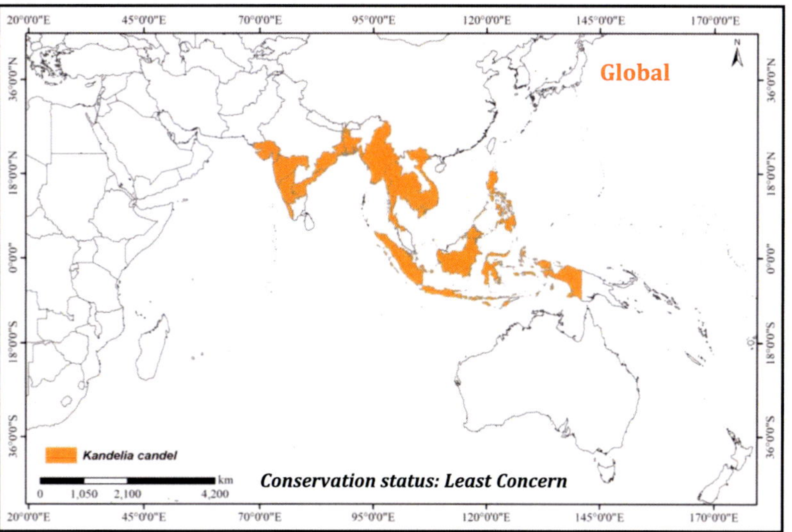

Habitat and Ecology

Often fringing the banks of mid estuarine position and also present as understory in mid intertidal position in association with *Avicennia alba*, *Sonneratia apetala.*

Phenology

Flowering and fruiting occur throughout the year. Flowering peaks from June to August, Fruiting from September to October.

Notes

In the past *K. obovata* of the coast of South China sea is identified as *K. candel*. However, they are two well differentiated geographical sets of populations in the genus *Kandelia* as evident by recent conventional and molecular taxonomic studies. Petal features, bracteole shape, and style length in relation to stamens are the key characters used for the distinction (Sheue et al. 2003).

Morphological features of *Kandelia candel*. (A) Habit; (B) leaves; (C) branchlet with leaves and flowers ; (D) mature flower bud; (E) opened flower; (F) stamens; (G) deeply bifid petal with bristles; (H) propagule

Leaves of *Aegiceras corniculatum*

Genus: *Lumnitzera*

Combretaceae

Lumnitzera belongs to the family Combretaceae represented by two species viz., *L. littorea* and *L. racemosa*, and one natural hybrid viz., *L. × rosea*, in global mangroves. Two species, viz., *Lumnitzera littorea* and *Lumnitzera racemosa* are known from Indian mangroves, of which, *L. littorea* is restricted only to A & N Islands whereas *L. racemosa* is commonly distributed in mainland India as well as in A & N Islands. Both the species have similar appearance, but they differ distinctly in flower colour. *Lumnitzera littorea* has red flowers, while *L. racemosa* has white flowers.

White flowers of *L. racemosa*

Key to *Lumnitzera* species

1. Inflorescence axillary, style central, petals white, stamens equal or slightly exceeding the petals............***Lumnitzera racemosa***
2. Inflorescence terminal, style off-centre, petals red, stamens exceed the petals........***Lumnitzera littorea***

Flower buds of *L. littorea*

Above ground roots of *L. littorea*

Leaves with emarginated tip

Lumnitzera littorea (Jack) Voigt, Hort. Suburb. Calcutt. 39 (1845).

Combretaceae

Lumnitzera littorea is most common in A & N Islands, but it is not known from east and west coasts of mainland India. Generally, it is found in landward edges of mangroves in association with *Ceriops tagal*, *Rhizophora* spp and *Bruguiera* spp and rarely occurs as monotypic stands. *Lumnitzera littorea* is distinguished from *L. racemosa* by its red colour flowers, terminal inflorescences, long stamens and eccentric style placement.

Succulent leaf with notched tip

KEY CHARACTERS

- Medium sized tree, but can grow up to 15m height
- Bark greyish brown and fissured
- Above ground roots are not usually present. However, in moist environments small knee like roots and cable roots may be visible
- Leaves are small light green in color, fleshy with an indentation at the tip
- Inflorescences terminal, spicate
- Flowers are red in colour, style exceeds the corolla
- Fruits yellowish green, glossy, vase- shaped, longitudinally ribbed

Species feature

Red color flower with long stamens

Yellowish green fruits

Fissured brown color bark

Lumnitzera littorea (Jack) Voigt

Taxonomy

Tree: spreading, height to 20 m, evergreen, multi-stemmed (A). *Bark*: dark brown, deeply fissured and flaky (G); twigs smooth, green becoming brown. *Roots*: wiry knee, looped above ground, stem simple. *Leaves*: simple, alternate, succulent, green-dark green (B), narrowly obovate, apex rounded and emarginate (C), base attenuate, 5–8 × 1.5–2.5 cm, ratio of length to width c.2; petiole rounded, green 0.5 cm long. *Inflorescence*: terminal, 5–15 flowered, 2–4 cm long (E); *Mature flower buds*: ellipsoidal (D), 1.5–2 cm long; peduncle short; calyx tube like, slender, green (H), 0.8–1.5 cm long, calyx lobes 5, apex pointed, 0.2 cm long; petals 5, bright red, 0.3–0.5 cm long, erect; stamens 10, longer than petals (H), 1–1.5 cm long; style simple, placed off-centered, 0.5–0.7 cm long. *Mature fruits*: clustered, drupe shaped (F), one seeded, 1–1.5 × 0.3–0.5 cm, style and calyx persistent, epicarp fibrous.

Distribution

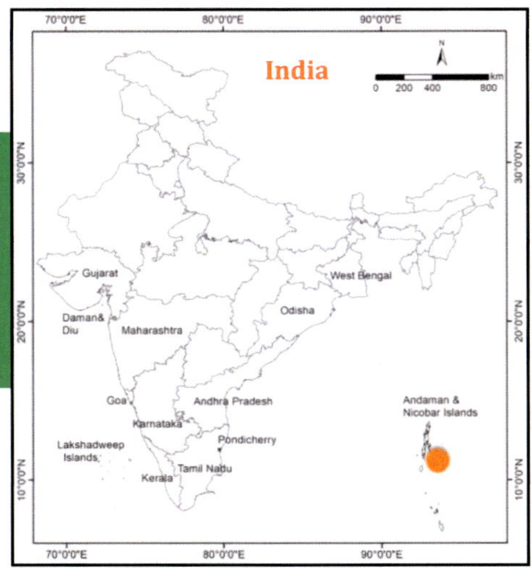

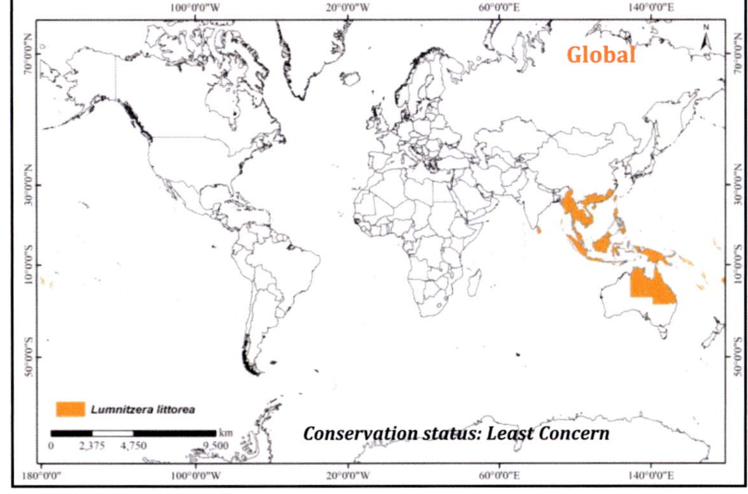

Habitat and Ecology

Often found in mid intertidal and intermediate estuarine position on soft muddy substrate at landward margin of mangroves flooded only during spring tides.

Phenology

Flowering and fruiting occur throughout the year. Flowering peaks from October to December; Fruiting from January to March.

Notes

Restricted distribution of *Lumnitzera littorea* to A & N Islands and Sri Lanka in Indian subcontinent remains unexplainable. Vivipary or cryptovivipary is not observed in *Lumnitzera* spp. Despite the high fruit setting, high percentage of mature fruits is commonly empty due to abortion of embryos. Sometimes embryos are also eaten by larvae that emerge from the eggs laid by insects in the early stages of fruit development.

Morphological features of *Lumnitzera littorea*. (A) Habit; (B) succulent leaves; (C) leafy rosette with young inflorescences; (D) young buds; (E) opened flowers; (F) fruits; (G) fissured bark; (H) flowers with long stamens and style.

Lumnitzera racemosa Willd., Neue Schriften Ges. Naturf. Freunde Berlin iv. 187 (1803).

Combretaceae

Lumnitzera racemosa is common in Indian mangroves. It is a shrubby tree or tree growing in land ward edges of mangroves and often occurs as monotypic stands. *Lumnitzera racemosa* is identified generally by its light green, relatively sparse foliage, and dark roughly fissured stems. It is distinguished from its near relative, *L. littorea*, by its axillary racemes of flowers with large, reflexed, white petals nearly as long as stamens, and its centrally placed style.

Fruits

KEY CHARCTERS

- Medium sized tree, but can grow up to 10m height
- Bark reddish brown
- Above ground roots are not usually present. However, in moist environments small knee like roots may be present
- Leaves are succulent spatula like with an apical notch
- Inflorescences axial and spicate
- Flowers are white in colour, style does not exceed the corolla
- Vase shaped fruits

Species feature

Light green leaves with apical notch

White color flowers with small stamens

Axillary racemes with vase shaped fruits

Lumnitzera racemosa Willd.

Taxonomy

Tree: spreading, height to 15 m, evergreen, multi-stemmed (A). *Bark*: grey, fissured and flaky (B), twigs smooth, green becoming brown. *Roots*: wiry knee, looped above ground, stem base simple (C), occasionally short buttresses. *Leaves*: simple, alternate, succulent, wavy margin, light green, narrowly obovate, apex rounded and emarginated (D), base attenuate 4.5–10 × 1.5–3.5 cm, ratio of length to width c. 3; petiole rounded 0.3–0.5 cm long. *Inflorescence*: axillary, 1–8 flowered, 2–3 cm long (F). *Mature buds*: ellipsoidal, 1–1.5 cm long; calyx lobes 5, apex pointed, 0.1 cm long, calyx tube green (E), 0.5–0.8 cm long; petals 5, white, elliptic, apex acute, 0.3–0.5 × 0.1–0.2 cm, reflexed; stamens 10, 2 enclosed by each petal, length equal to petals (E); style 0.4–0.5 cm(G, H); *Mature fruits*: Vase shaped, yellowish green, 1–1.5 × 0.5–0.6 cm, one seeded, style and calyx lobes persistent, epicarp fibrous.

Distribution

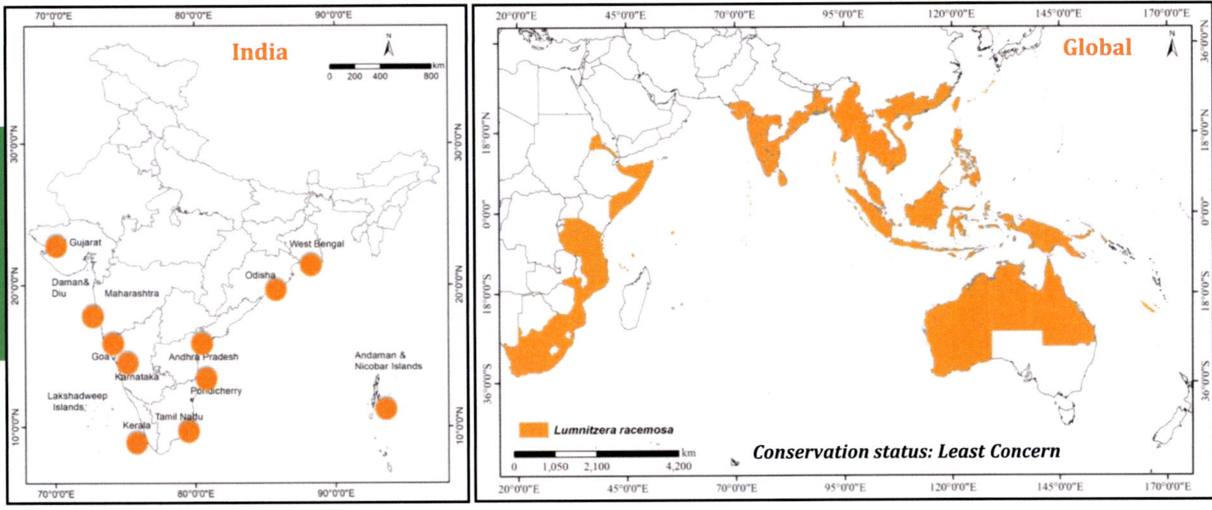

Habitat and Ecology

Mostly found in mid and high intertidal and intermediate estuarine position on the landward fringe sandy/ muddy substratum.

Phenology

Flowering and fruiting occur throughout the year. Flowering peaks from October to December; Fruiting from January to March.

Notes

In *Luminitzera racemosa* also, natural regeneration is low due to high percentage of embryo abortion. *Lumnitzera × rosea* is a putative hybrid between *L. littorea* and *L. racemosa*, and this hybrid is distinguished from its parents by its terminal racemes of flowers with long, reflexed pink petals nearly as long as the stamens and slightly eccentric style placement (Duke 2006). Despite the overlapping distribution of parental species, *L. × rosea* is not known from A & N Islands.

Morphological features of *Lumnitzera racemosa*. (A) Habit; (B) bark; (C) stem base; (D) leaves; (E) flower with stamens equal to petal length; (F) axillary inflorescences; (G &H) drupe shaped fruits.

Nypa fruticans Wurmb, Verh. Batav. Genootsch. Kunst. i. 349 (1779). Arecaceae

Nypa fruticans is the primitive as well as the only mangrove species in the genus *Nypa* belonging to the family Arecaceae, commonly called as "Mangrove Palm". *Nypa fruticans* is distinguished from other palms by its rhizomatous, under growth with dichotomous branching that facilitates asexual reproduction. It has a large densely packed, globose fruiting heads with numerous fibrous fruits. It is commonly distributed in A & N Islands, whereas in mainland India it has restricted distribution in Sundarbans. It has also been introduced in few areas of Kerala and Maharashtra in recent times.

KEY CHARACTERS

- Palm growing in low saline, sheltered inter-tidal creeks
- Aerial root absent
- Leaves are lanceolate, palm leaf, arising from root stock, yellow when young leaflet tip acute
- Flowers-female in globose head, male in catkin like, brick red to yellow
- Fruit -dark brown or brick red, globose, single seeded carpel, pericarp fleshy, fibrous

Species feature

Habit

Subterranean stem

inflorescences

Large spherical fruiting head

Nypa fruticans Wurmb

Taxonomy

Palm: stem less, height to 10 m, evergreen (A). *Root*: rhizomatous, dichotomously branched below ground (G). *Leaves*: erect, slightly recurved, 4–8 m long, alternate, paripinnate compound, oblong to ovate elliptic, leaf stalk stout, 1–1.4 m long, strongly flanged at the base, enclosing the stem; leaflets 30–40 pairs, lanceolate with upper green surface, and powdery lower surface, 1–1.5 m long, apex narrowly pointed, midrib prominent on adaxial. *Inflorescence*: axillary, bisexual (B). *Mature flowers*: borne on long sturdy peduncles consist of main axis and several lateral branches; female inflorescences found on main axis as globose cluster of congested flowers (E), flowers with 6 calyx lobes, 0.4–0.5 cm long, stigma sessile, funnel shaped; male inflorescences found on lateral axis as cub-shaped spike of densely packed flowers (F); calyx lobes 6, 0.4–0.5 cm long; stamens 3, united as central column. *Mature fruit*: fruiting body is an aggregate of 30–50 densely packed fertile and sterile carpels (D), spherical in shape, 30–40 cm in diameter; individual carpel is one seeded (C), smooth, dark brown, distal end bulbous, 5–10 × 3–6 cm, pericarp fleshy, fibrous, endocarp spongy; seed egg shaped, 4–5 cm long, germination incipiently viviparous.

Distribution

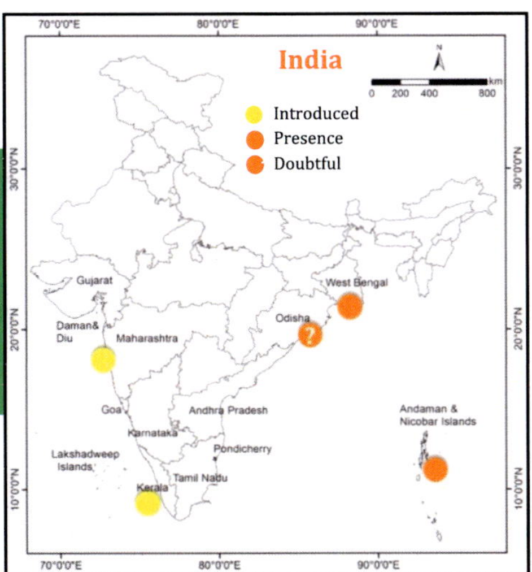

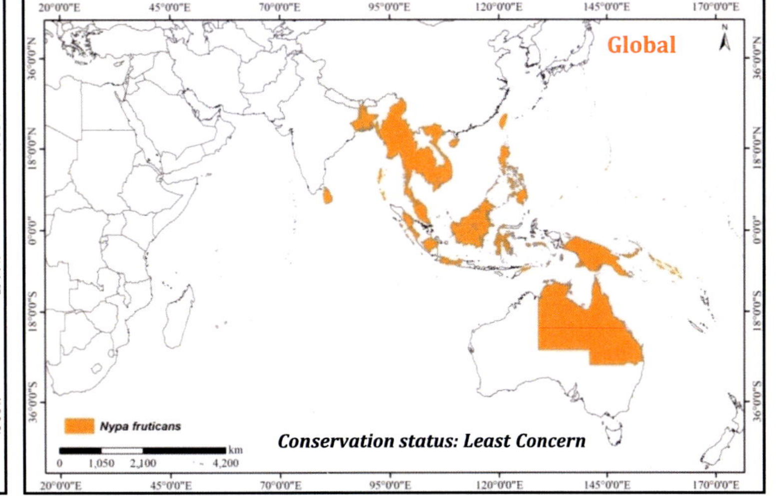

Habitat and Ecology

Often found in upstream estuarine position on soft, fine-grained substrates. It usually occurs in pure stands where plenty of freshwater input is available.

Phenology

Flowering and fruiting occur throughout the year.

Notes

Nypa fruticans was listed from Malabar Coast by Van Rheede during 17th century itself (1678-1693) in his classical work "Hortus Malabaricus". So, the presence of this species along the west coast of India needs to be validated through undertaking exclusive surveys.

Morphological features of *Nypa fruticans*. (A) Habit; (B) inflorescences; (C) individual carpel with seedling; (D) fruits with aggregate of carpels; (E) female inflorescences; (F) male inflorescences; (G) rhizome.

Pemphis acidula J.R.Forst. & G.Forst., Char. Gen. Pl. 34 (1775). Lythraceae

Pemphis acidula is the only mangrove species in the genus *Pemphis* belonging to the family Lythraceae in IWP region. It is typically the mangroves of coral reef ramparts (Duke 2006). In India it is present in the coast of Tamil Nadu, Lakshadweep and A & N Islands. In the field, it can be easily recognized by its small pubescent leaves. *Pemphis acidula* has two distinct forms of flowers with characteristic crumbled petals. Both forms have both male and female parts. One form is called 'thrum' having short style ending with stamens, while other is called 'pin' having long style (Duke 2006). Generally, it is found in downstream of the intertidal mangrove habitat.

Species feature

Swollen mature fruits capsules

KEY CHARACTERS

- Tree/shrub growing in the apparent sandy/rocky mangrove shore
- No prominent aerial root
- Bark light grey to brown
- Leaves very small, fleshy, simple, opposite, elliptical to obovate, tip rounded to bluntly acute
- Flower- axillary, 1 to few flowered cymes; calyx 12 lobed, green; corolla -6, white; stamens up to 12
- Fruits hairy, green, 30 small seeds in fruits.

Habit

Small pubescent leaves

Short styled thrum flower

Style

Pemphis acidula J.R. Forst. & G. Forst.

Taxonomy

Tree: medium sized tree, occasionally shrub, height to 5 m, multi stemmed, and evergreen (A). *Twigs*: green turns brown on maturity, densely hairy. *Bark*: pale brown, deeply fissured and shedding in long curling strips (E). *Roots*: occasionally above ground, stem base simple. *Leaves*: simple, opposite, pale green, hairy (B), 1–3 × 0.5–1 cm, margin entire, apex rounded to obtuse pointed, base cuneate; petiole pale green, hairy, 0.1–0.2 cm long. *Inflorescences*: axillary (D, H), solitary, rarely 2–3 flowered, bracteolate sessile cymes, peduncle 0.5–1 cm long, hairy. *Mature flowers*: perfect, greenish- red, campanulate, calyx 0.5–0.8 cm long, 12-ribbed turbinate tube with 6 erect-pointed triangular lobes (C), 0.1–0.2 cm long; calyx lobes valvate initially closing calyx cup, redden after opening; petals 6, white, elliptic, inserted at mouth of calyx cup below sinus (G), 0.5–0.8 cm long; stamens 12, in 2 series inserted on lower third of calyx cup (C, G), 0.2–0.3 and 0.4–0.5 cm long; style simple, stigma capitate, dark green, based on style length two forms are recognized 'thrum' form has short style (I), 0.1 cm long, stamens enclosing stigma; 'pin' form has long style (D), up to 0.4 cm long, stigma exceeding stamens. *Mature Fruit*: spherical capsule, reddish-green, enclosed in enlarged obovate calyx tube (I), 0.4–0.5 cm wide, style persistent; seeds numerous 20–30, angular, 0.2–0.3 cm long.

Distribution

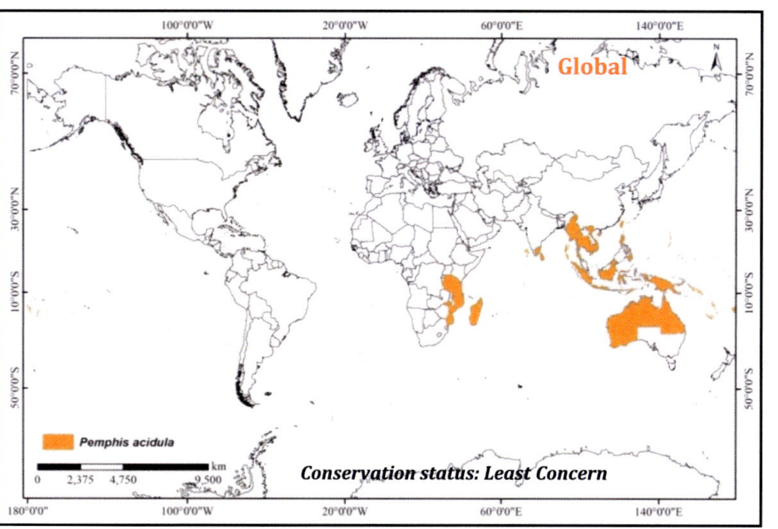

Morphological features of *Pemphis acidula*. (A) Habit; (B) small pubescent leaves; (C) calyx lobes; (D) leafy rosette with fruit; (E) bark (F) young bud; (G) flower with white petals; (H) calyx tube; (I) fruit.

Habitat and Ecology

Found in landward margin of mangroves, often above the high tide level, and on beaches. Occur on sand, laterite, limestone and gravel.

Phenology

Flowering and fruiting occur throughout a year. Flowering peaks from February to April; Fruiting from May to August.

Notes

Pemphis acidula is often included as mangrove associate. This taxon is also considered to be an intermediate between a strand plant and a mangrove (Tomlinson 1986). However, *Pemphis acidula* is dominant constituent of mangroves areas dominated by coral deposits in Gulf of Mannar, Lakshadweep Islands, Palk bay, Havelock and Neil islands of A & N Islands.

Phoenix paludosa Roxb., Hort. Bengal. 73; Fl. Ind. iii. 789 (1832). Arecaceae

Phoenix paludosa belonging to the family Arecaceae is known to be found in Sundarbans, Odisha and Andaman Islands and mostly occur towards the landward margin of the mangroves in association with *Cynometra iripa*. However, in Andaman Islands it can grows in the banks of mid and upper tidal creek along with *Xylocarpus* spp., and *Rhizophora* spp. In the field it is easily recognized by it palm like leaves.

Immature fruits

KEY CHARACTERS

- Palm, up to 6 m tall, but often much shorter
- Leaves are palm like
- The unisexual flower cluster looks like a stiff brush, and is located between the leaves on an erect stalk of 60 cm
- The orange berries are oval, and about 1 cm long. The seeds have a lateral embryo, unique in *Phoenix*.

Species feature

Orange color berry like fruits

Leaves **Above ground roots**

Inflorescences

Phoenix paludosa Roxb

Taxonomy

Palm: height to 6 m, but often much shorter (A). *Roots*: thin pneumatophores lie above ground root (D). *Stem*: slender, grey, have persistent spiny leaf stalks, fibrous leaf sheaths and diamond-shaped leaf scars (C). *Leaves*: imparipinnate, numerous bright green (B), sometimes yellowish persist near the top of the stem and are pale grey underneath, the leaves curve slightly, are inserted on the trunk so that they point upward, and are rather short, measuring up to 150 × 45 cm, midrib strong, ending into strong sharp spine at apex few pairs of lower segments modified into sharp spines. *Inflorescence* a spadix; spadices thickly coriaceous, simple branched, arising in between leaves; spathes 20–30 cm long, brownish, enclosing the flowers. *Mature flowers*: Flowers dioecious, yellowish-white, small, tri-merous; stamens 3 (in male flowers); carpels 3 (in female flowers). *Mature fruits*: the orange berries are oval (E-G), and about 1 cm long; seeds have a lateral embryo, unique in *Phoenix*.

Distribution

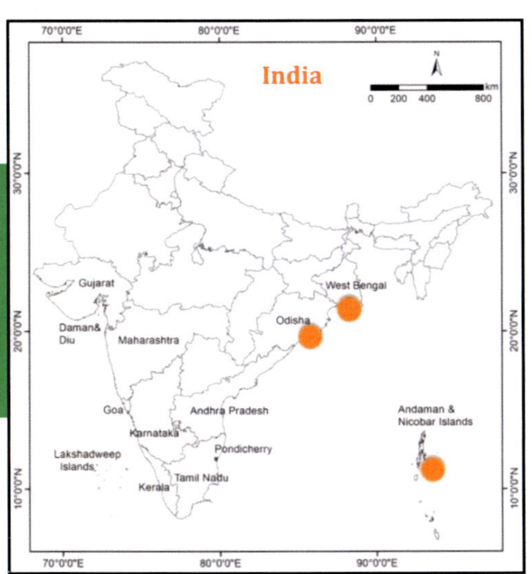

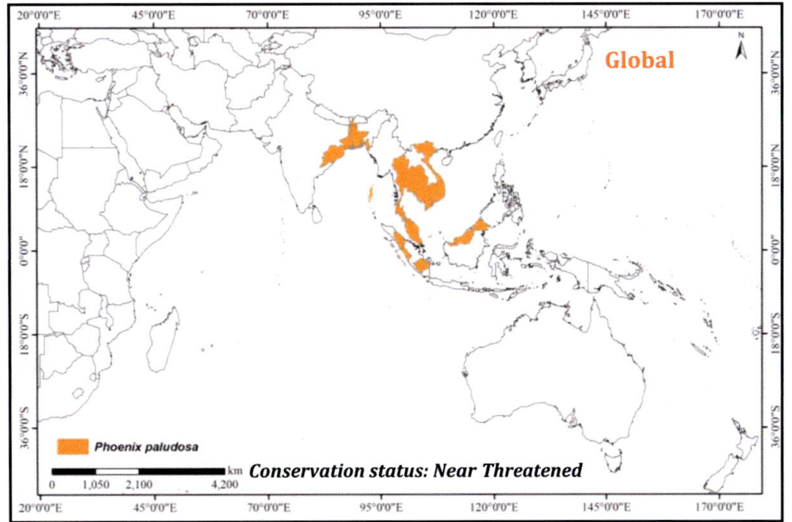

Habitat and Ecology

Often occur on the landward margin of mangroves, forming extensive, dense, leafy clumps with stems of all heights

Phenology

Flowering February to April; Fruiting May to July.

Notes

Phonix paludosa is often termed as mangrove associate. However, it possesses adaptive features such as thin pneumatophores, and water dispersed seeds, to grow in the mangrove niche.

Morphological features of *Phoenix paludosa*. (A) Habit; (B) leaves; (C) stem; (D) above ground roots; (E) fruit with calyx; (F) cluster of fruits; (G) ripened fruits.

Stilt root of *Rhizophora stylosa*

Genus: *Rhizophora*

Rhizophora is a key member of family Rhizophoraceae. *Rhizophora* is distinguished from other genera by its prominent stilt root, leaves with spiked mucronate tip, dark spots on the underside of leaves, flowers with four-pointed calyx lobes and separate fruiting body and viviparous propagules. Species of Genus *Rhizophora* are divided into two broad regional groupings i.e., IWP stilt mangroves (*R. apiculata*, *R. mucronata*, *R. stylosa*, *R. ×annamalayana*, *R. ×lamarckii*) and the AEP red mangroves (*R. mangle*, *R. racemosa*, *R. selala*). IWP species are readily distinguished from red AEP species by prominent spiked, mucronate tip at leaf apex, instead of a blunt recurved tip (Duke 2006).

Key to *Rhizophora* species

1. (a) Peduncle shorter than petiole..2
   (b) Peduncle as long as or longer than petiole 4
2. (a) Mature flower bud and fruits below the leaves; inflorescences two-flowered; bract corky, brown; hypocotyls present..................................... Rhizophora apiculata.
2. (b) Mature flower buds within the leaves; inflorescences 2-4-flowered; bract smooth and green; hypocotyls not present...3
3 (a) Leaves broadly elliptical; styles 0.8–1.2 mm long; stamens in two whorls, inner shorter; mature flower bud four-sided in cross-section....... ***Rhizophora ×annamalayana***
   (b) Leaves narrowly elliptical; styles 2–3 mm long; stamens in one whorl; mature flower bud rounded in cross- section..............................***Rhizophora ×lamarckii***
4. (a) Mature bud < 1.5cm long..............5
   (b) Mature bud >1.5 cm long...........7
5. (a) Stamens 4, petals thick and leathery, densely hairy..........*Rhizophora mucronata* var. *alokii*
   (b) Stamens 8; petals thin, hairy at margin...6
6. (a) Bract and bracteoles minute; style <1 mm long, seated on elongate, tapering ovary; hypocotyls 50–80 cm long..................***Rhizophora mucronata***
   (b) Bract and bracteoles prominent, forming two-lobed, cup-like structure; style 3-4 mm (>2.5mm, seated on short ovary; hypocotyls 20–40 cm long.............***Rhizophora stylosa***
7. (a)Leaves oblong, apex rounded, peduncle longer than pedicle, style 0.4-0.5 cm, hypocotyls absent................ ***Rhizophora ×mohanii***
   (b) Leaves elliptic to broadly elliptic, apex acute, pedicel longer than peduncle with style 0.2-0.25 cm, hypocotyls present ....***Rhizophora stylosa*** var. ***andamanica***

Rhizophora apiculata Blume, Enum. Pl. Javae 1: 91 (1827).

Rhizophoraceae

Rhizophora apiculata one of the dominant mangrove species of IWP. In India *R. apiculata* is abundant in A & N Islands, while in east and west coast of mainland India its distribution is patchy. Generally found in middle and upstream areas of the tidal creek along with *R. mucronata, R. stylosa, Bruguiera gymnorhiza* and *Xylocarpus* species. *Rhizophora apiculata* is readily distinguished from other *Rhizophora* species by its narrow, apiculate glossy dark green leaves, inflorescences with short peduncle and flower with corky brown bract.

Propagule

KEY CHARACTERS

- Tree grows up to 10–15m height, stilt roots and prop roots present
- Bark- rough, brown to dark grey, slightly fissured
- Leaves dark green, narrowly elliptic with acute leaf tip and prominent mucronate, dark spots on underside present
- Inflorescences axial, peduncle short (<1cm)
- Flowers oval shaped, bract corky brown, calyx woody, 4 lobed,
- Petals white, linear, smooth, stamens 9–12, style bi-lobed obscure
- Propagules 20–40cm in length

Narrow acute leaves

Species feature

Bud with corky brown bract

Bifurcated small style

Flowers with petals and Stamens

Stilt root

Rhizophora apiculata Blume

Taxonomy

Tree: columnar, height to 25 m, evergreen (A). *Bark*: dark grey rough, slightly fissured, looks like crocodile skin (C). *Roots*: both stilt roots and aerial roots growing from lower branches, stilt roots are highly conspicuous arching above ground to 3 m (B). *Leaves*: simple, opposite, green-dark green in colour (D), margin entire, 8.5–16.5 × 4–8.5 cm, ratio of length to width c. 2.0 (not less than 2cm), base cuneate, apex acute with pointed mucronate tip (E), 0.4–0.5 cm long; petiole smooth, green or reddish, 1–2.5 × 0.2–0.3cm. *Inflorescence*: axillary, 2 flowered (F); peduncles 0.8–1 × 0.4–0.6 cm; bract short, connate; bracteoles beneath the calyx swollen corky brown (F). *Mature flower buds*: broadly elliptic, yellowish brown, often fissured on maturity, 1–1.6 × 1 cm, ratio of length to width c. 1.2, apex acute, cross section circular (H); calyx lobes 4, thick, apex acute; petals 4, greenish or creamy white (G), smooth without pubescence, thin, 0.7–1 × 0.2 cm, not enclosing the stamens, flat; stamens 9–12 (up to 16), pale brown (G), 0.8–1.1 cm long; style not more than 0.1–0.12 cm, seated on domed ovary (I). *Mature fruit*: inverted pear shaped, brown, 2–3 × 1.5–2.5 cm, calyx persistent, lobe erect (J). *Mature hypocotyls*: green, smooth, 20–40 cm long, 1–1.5 cm width at the widest point, tip bluntly pointed; plumule green, 1.5–2.5 cm long (J).

Morphological features of *Rhizophora apiculata*. (A) Habit, tall form; (B) thicket form; (C) bark; (d) leafy rosette; (E) apex with mucronate; (F) mature bud; (G) flower; (H) bud cross section; (I) style; (J) mature propagules.

Distribution

India

Presence
Doubtful

Gujarat, Daman& Diu, Maharashtra, Goa, Karnataka, Kerala, Lakshadweep Islands, Tamil Nadu, Pondicherry, Andhra Pradesh, Odisha, West Bengal, Andaman & Nicobar Islands

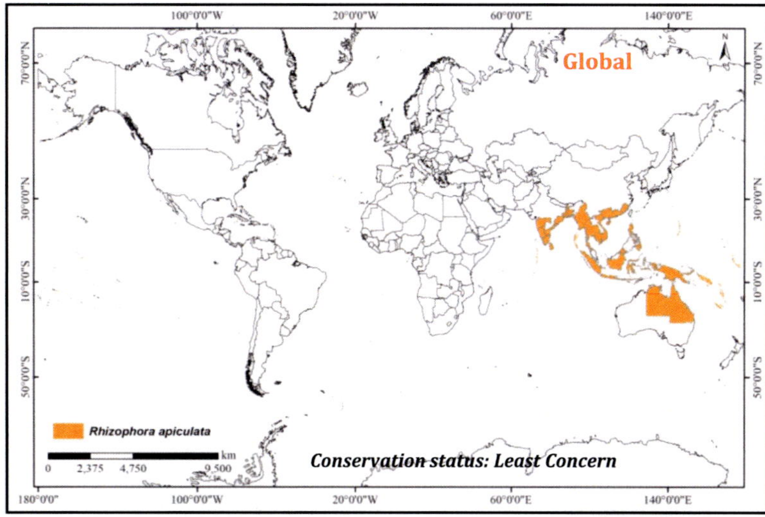

Habitat and Ecology

Often found in mid-intertidal and intermediate estuary. Individuals growing on muddy soil in intermediate tidal creeks are taller whereas individuals at mouth of the creek or upstream areas are small in nature forming thickets.

Phenology

Flowering and fruiting occur throughout the year. Flowering peaks from January to April; Fruiting from August to November.

Notes

Identity of *R. apiculata* is most consistent in the genus *Rhizophora*. However, *R. apiculata* exhibits different growth forms depending on the substrate or position along the creek. Individuals fringing the bank of the creeks are short compared to those present in inner mangrove areas. Similarly, individuals on sandy and rocky substrate are short and bushy, while those on muddy substrate are taller.

Rhizophora mucronata Lam., Encycl. 6: 189 (1804).

Rhizophoraceae

Rhizophora mucronata is another dominant mangrove species of the IWP as well as India. *Rhizophora mucronata* is readily distinguished from *R. apiculata* by multi-flowered inflorescences, long peduncle, broadly elliptic leaves. Style length is the main feature used to differentiate this taxon from *R. stylosa* viz., <0.25 cm in *R. mucronata*, while > 0.25 cm in *R. stylosa* (Duke 2006).

Long propagules

KEY CHARACTERS

- Tree grows up to 10–15m height, stilt root extensively developed
- Bark rough, brown to dark grey, fissured with thick flakes
- Leaves light green, broadly elliptical, plenty of dark spots on underside of leaves, mucronate prominent
- Inflorescences axial, peduncle long, bract and bracteoles obscure, 2-8 flowers
- Flowers oval shaped, calyx four lobed, petals 4, hairy, stamens 8, style obscure <0.1 cm seated on raised ovary.
- Hypocotyls up to 80cm long, tapering towards end.

Species feature

Yellowish green broad leaves

Small style with raised ovary

Inflorescences with long peduncle

Open flower with small style and hairy petal

Rhizophora mucronata Lam.

Taxonomy

Tree: columnar to sprawling, height up to 20 m, evergreen (A). *Bark*: dark grey, rough, fissured with thick flakes (J). *Roots*: both stilt roots and aerial roots growing from lower branches (A). *Leaves*: simple, opposite, green-yellowish green in colour, broadly elliptic to oblong ovate (B), underside with plenty of dark spots, 8.5–15 × 5.7–9 cm, length to width ratio averaging 1.6 (not greater than 1.8), apex acute to obtuse with mucronate spike, 0.4–0.5 cm long (K), base rounded or slightly cuneate, margin entire; petiole green 1.5–3 × 0.3–0.5 cm. *Inflorescences:* axillary cymose (C), mostly 2–8 flowers rarely exceeds 8; bract and bracteoles obscure (F, D); peduncle 1.5–6 × 0.3–0.5 cm. *Mature flower bud*: ovate, wider towards base, apex acute to obtuse (D), cross section rounded (E), 1.2–1.6 × 0.8–1, length to width ratio ca. 1.8; calyx 4 lobed, thin, yellowish white; petals 4, creamy white, thin, hairy (G), often enclosing stamens, 0.8–1 × 0.3–0.4 cm; stamens 8, pale brown, 0.7–0.9 cm long (G); style terete with bilobed stigma, mounted on raised ovary (H), mostly 0.8–0.12 cm long. *Mature fruit*: pear shaped, 4–4.5 × 2.5–3 cm, calyx persistent and lobes erect. *Mature hypocotyls:* green, 50–80 cm long (I), 1.5–1.7 cm wide at the widest point, tip narrowly pointed; plumule green, 2–3 cm long.

Distribution

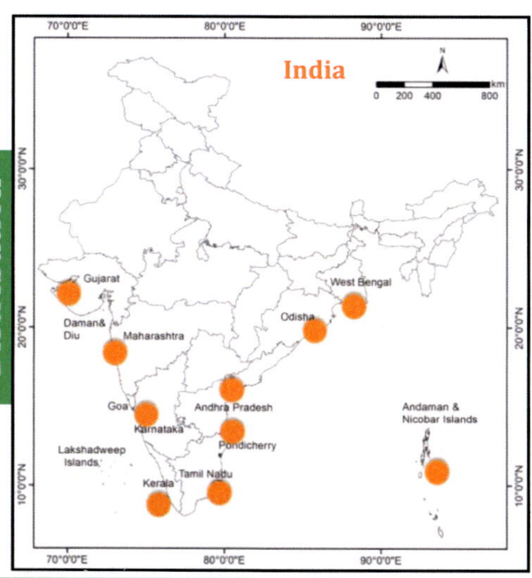

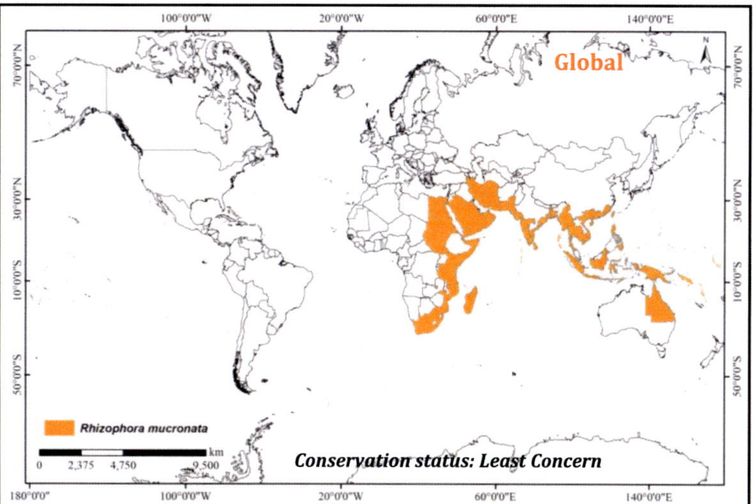

Habitat and Ecology

Found commonly in intermediate to upstream tidal creeks and low intertidal position. Generally, grows in clusters near or on the banks of creeks and tidal estuaries, rarely far from the tides.

Phenology

Flowering and fruiting occur throughout the year. Flowering peaks from September to December; Fruiting from January to April.

Notes

Identity of *R. mucronata* is most consistent, if it occurs in association with *R. apiculata*, whereas its identity is often confusing when it occurs in association with *R. stylosa*. Both *R. mucronata* and *R. stylosa* have 2-8 flowers and longer peduncle than petiole. However, leaves with rounded base, large sized bud (>1.2cm long), minute bract and bracteoles, short style (<1 mm) seated on raised ovary, and long hypocotyls (up to 80 cm) are the key diagnostic features of *R. mucronata*.

Morphological features of *Rhizophora mucronata*. (A) Habit; (B) leaf rosette with flowers; (C) inflorescences; (D) ovate mature bud; (E) mature bud cross section rounded; (F) minute bract at dichotomous branch; (G) flower with hairy petals; (H) style seated on raised ovary; (I) mature propagules; (J) bark; (K) mucronate tip.

Rhizophora stylosa Griff., Not. Pl. Asiat. 4: 665 (1854). Rhizophoraceae

In India *Rhizophora stylosa* is restricted only to A & N Islands. In A & N Islands *R. stylosa* was first reported by Mall et al. (1985) from Burmanallah, Chidiytaapu and Corbyn's cove. Subsequently *R. stylosa* was included in the mangrove floral list of A & N Islands by many without locality data. Ragavan (2015) reported the occurrence of *R. stylosa* in 10 localities in A & N Islands. *Rhizophora stylosa* is easily distinguished from *R. mucronata* by its relatively long style, small flowers, short propagules (<40 cm) and dark green slightly broad leaves.

Mature propagules

KEY CHARACTERS

- Tree grows upto 5–10m height, possess extensively developed stilt root and prop root
- Bark- rough, brown to dark grey, fissured with flakes
- Leaves dark green, plenty of dark spots on underside
- Inflorescence axial, peduncle long (longer than petiole length), 2–14 flowered
- Flower oval shaped, smaller than flowers of *R. mucronata*
- Calyx four lobed, petals 4, hairy, stamens 8, style 0.2–0.4 cm, ovary obscure
- Propagules less than 40 cm, tapering towards end

Dark green leaves.

Inflorescences with long peduncle

Flowers with hairy petals

Species feature

Long style <2.5mm

Rhizophora stylosa Griff.

Taxonomy

Tree: spreading, height to 10m, evergreen (A). *Bark:* smooth grey to dark grey, finely fissured (M). *Roots*: both stilt roots and aerial roots growing from lower branches (A). *Leaves:* simple, opposite, green-dark green, oblong elliptic to narrowly obovate (B), broader at apex (C), 8–12 × 4–6.5 cm, length to width ratio averaging 2 (not less than 1.8), apex acute to bluntly pointed with mucronate tip, 0.4–0.6 cm long, base cuneate, margin entire; petiole green, 2–3.5 × 0.3–0.4 cm. *Inflorescences*: axillary cymose, 2–8 flowered rarely exceeds 8 (E); bract and bracteoles prominent, two lobed, forms cup like structure (J); peduncle 2.5–5.5 × 0.2–0.3 cm; pedicels 0.5–1 cm long. *Mature bud*: ellipsoid, wider towards base (F), 0.7–1.2 × 0.3–0.6 cm, length to width ratio ca.2.39, cross section rounded (G); calyx 4 lobed, thin, yellowish white; petals 4, creamy white, thin, hairy (H), often enclosing stamens 0.7–0.9 × 0.2–0.4 cm; stamens 8, pale brown, 0.4–0.6 cm long; style terete with bilobed stigma, 0.3–0.5 cm long (I); ovary is not prominent. *Mature fruit*: pear shaped, dark brown (K), smooth, 2.4-2.7 × 1.9–2.3 cm, calyx persistent with erect lobes. *Mature hypocotyls*: green, 21–35 cm long, 1.3–1.6 cm wide and widest point, tip narrowly pointed (L); plumule green, 1.5–2 cm long.

Distribution

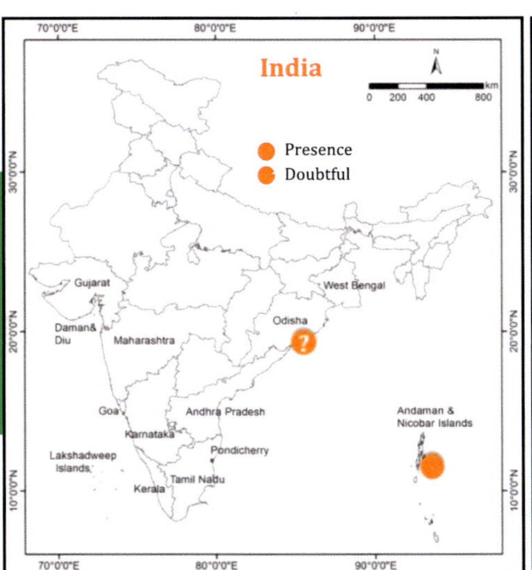

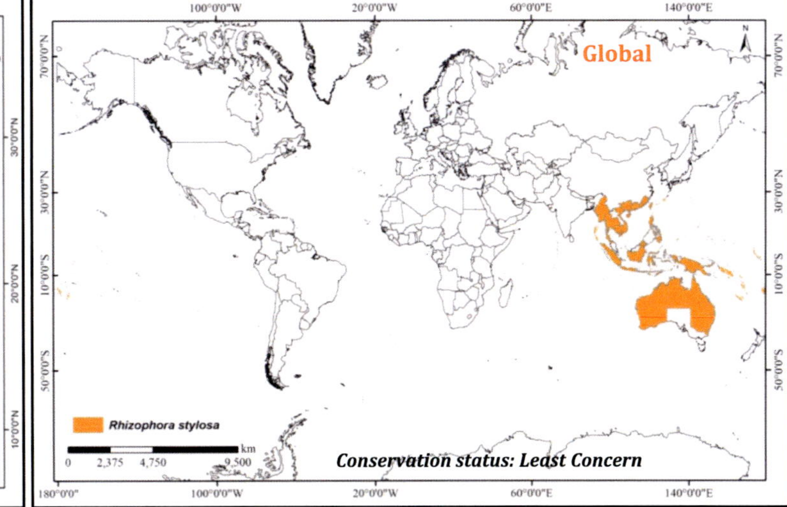

Morphological features of *Rhizophora stylosa*. (A) Habit; (B) leafy rosette with flowers; (C) narrowly obovate leaf (broader at apex); (D) mucronate; (E) inflorescences; (F) mature bud; (G) cross section slightly four sided; (H) flower; (I) style; (J) prominent bract at dichotomous branch; (K) fruit; (L) mature propagules; (M) finely fissured bark

Habitat and Ecology

Often found in mid to low intertidal and downstream tidal creeks; grows in a variety of habitats and disrupted mangrove areas. One distinctive niche is its ability to grow on edges of small coral islands, establishing on the coral substrate.

Phenology

Flowering and Fruiting occur throughout the year.

Notes

The taxonomical distinction between *R. stylosa* and *R. mucronata* is a debated topic of the genus *Rhizophora*. *R. stylosa* is described as an ecological variant of *R. mucronata*. However, recent molecular studies reveal that existence of genetic distinction between them, despite frequent hybridization and introgression (Wee et al. 2015).

Rhizophora ×lamarckii Montrouz., Mém. Acad. Roy. Sci. Lyon, Sect. Sci. 10: 201 (1860).

Rhizophora × lamarckii is a putative hybrid between *R. apiculata* and *R. stylosa*. In India *R. x lamarckii* was first reported by Singh et al (1987) from Havelock Island. In the field, *Rhizophora* hybrids are easily identified by their height, a large number of flowers with smooth bract, swollen bracteoles and rare occurrence of propagule. *Rhizophora × lamarckii* is distinguished from *R. × annamalayana* by its narrow leaves (length: width ratio >1.8), and long style (> 0.15 cm) and stamens 8–16, usually in one single whorls.

Stamens in two whorls

KEY CHARACTERS

- Tree grows up to 10-15m, extensively developed stilt root
- Bark rough, brown to dark grey, fissured and flaky
- Leaves dark green, narrowly elliptic, acute leaf tip, plenty of dark spots on underside
- Inflorescences axial, 2–4 flowered, peduncle short, bract and bracteoles smooth, green
- Flowers oval shaped, creamy in color, calyx four lobed, petals 4, hairy, stamens 8-16, often in single whorls, can also be in two whorls, style >0.15 cm, bi-lobed, often tinged red
- Propagules rarely occur, often not found

Smaller stamens

Green narrow elliptic leaves

Two flowered inflorescences with short peduncle

Long style > 2mm

Species feature

Bark

Rhizophora × lamarckii Montrouz.

Taxonomy

Tree: spreading, multi-stemmed, height to 25 m, evergreen (A). *Bark*: dark grey rough friable, horizontally fissured, chequered (J). *Roots:* both stilt roots and aerial roots growing from lower branches, stilt roots are highly conspicuous arching above ground to 3m. *Leaves* simple, opposite, green-dark green, elliptic - narrowly elliptic (F), 8–16 × 4.5–8.5 cm, length to width ratio averaging 2 (not less than 1.8), apex acute with pointed mucronate tip to 0.4–0.5 cm long (B), base cuneate to attenuate, margin entire; petiole green occasionally red, 1–3 × 0.3–0.4 cm; *Inflorescences*: axillary, 2–4 flowered (C); peduncle 1–2.5 × 0.3–0.4 cm; pedicel stout; bract swollen, smooth, green (E); *Mature flower bud:* narrowly ovate-ellipsoidal, yellowish green (E), 1.5–1.7 × 0.7–0.8 cm, length to width ratio ca. 2.06 (not less than 1.8), cross section slightly four sided (D); calyx 4 lobed, yellowish white, thick, apex acute; petals 4, creamy white, thin, slightly hairy in the margin (G, H), linear, 0.9–1.1 × 0.2–0.3 cm; stamens 8–14, occurs in single or two distinct whorls (H, I), pale brown; style bilobed, 0.25–0.3 cm long, seated on domed ovary (G). *Mature fruits*: not observed.

Distribution

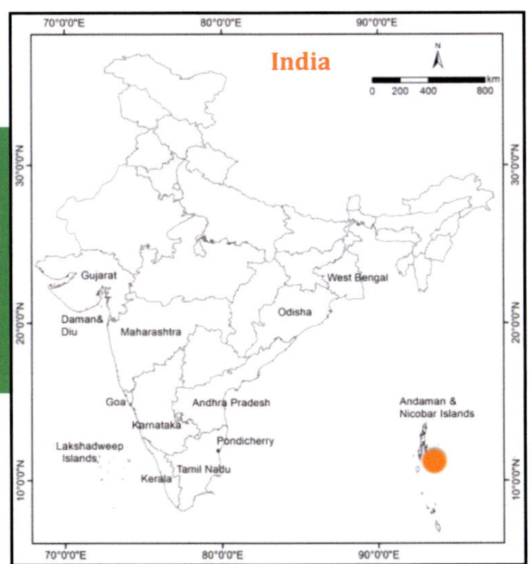

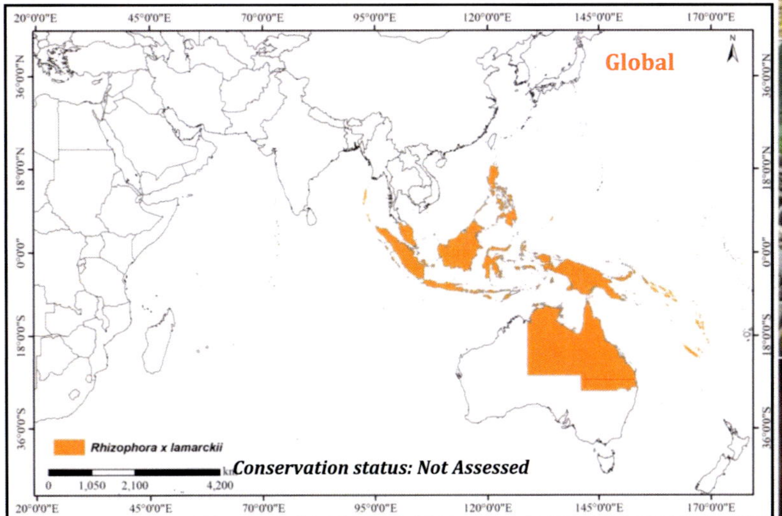

Habitat and Ecology

It is commonly found at mid intertidal areas and intermediate and downstream estuarine position in association with *R. apiculata* and *R. stylosa*.

Phenology

Flowering occurs throughout the year. Fruits are rarely present.

Notes

Rhizophora hybrids (*R. × lamarckii* and *R. × annamalayana*) are often resemble *R. apiculata*. However, *Rhizophora* hybrids are distinguished from *R. apiculata by* their smooth green bract and bracteoles and 2-4 flowered inflorescences within the leaf axils. Hybrid status and parentage of *R. ×lamarckii* are confirmed using molecular markers and hybridization was found to be bidirectional (Lo 2003, 2010; Ng et al. 2013). *Rhizophora ×lamarckii* is the best-known hybrid in *Rhizophora*.

Morphological features of *Rhizophora × lamarckii*. (A) Habit; (B) apiculate leaf tip with mucronate; (C) Inflorescences; (D) mature bud cross section four sided; (E) narrowly ovate mature bud; (F) leaf rosette with flowers; (G) style and slightly hairy petal; (H) inner smaller stamens; (I) flower; (J) bark.

Rhizophora × *annamalayana* Kathiresan, Environm. Ecol. 17(2): 500 (-501) (1999).

Rhizophoraceae

Rhizophora × annamalayana is a putative hybrid between *R. apiculata* and *R. mucronata*. In India *R. x annamalayana* was first reported by Kathiresan (1995) and considered as endemic to Pichavaram. But Ragavan et al. (2015d) reported the occurrence of *R. × annamalayana* in South Andaman, Middle Andaman, Mayabunder, Havelock and Carnicobar Islands of A & N Islands. *Rhizophora × annamalayana* is distinguished from *R. × lamarckii* by its relatively small style (<0.15 cm), occurrence of stamens in two rows i.e., outer 8 long stamens and inner 8 smaller stamens and broad acute leaves.

Propagule

KEY CHARACTERS

- Tree grows up to 10–15m, extensively developed stilt root
- Bark rough, brown to dark grey, fissured and flaky
- Leaves dark green, broadly elliptic with acute leaf tip, plenty of dark spots on underside
- Inflorescences axial, 2–4 flowered, peduncle short, bract and bracteoles smooth, green
- Flowers oval shaped, creamy in color, calyx four lobed, petals 4, hairy, stamens 8–16, often in two whorls (outer larger stamens, inner smaller stamens), style <0.15 cm, bi-lobed, seated on raised ovary
- Propagules rarely occur, if present similar to *R. apiculata*

Dark green broad leaves

Inflorescences with smooth green bract and short peduncle

Species feature

16 stamens (8 large and 8 small)

Small style < 1,5mm

Rhizophora × *annamalayana* Kathiresan

Taxonomy

Tree: spreading, multi-stemmed, height to 25 m, ever green (A). *Bark*: light grey, rough, slightly fissured both horizontally and vertically (K), resembles *R. apiculata*. *Roots*: extensively developed stilt roots and aerial roots growing from lower branches, stilt roots are highly conspicuous arching above ground to 3 m. *Leaves*: simple, opposite, broadly elliptic, yellowish green to dark green (B), 12–16 × 6–10 cm, length to width ratio averaging 1.67 (not greater than 1.8), base cuneate (C), apex acute with pointed mucronate to 0.5 cm long (O), margin entire; petiole thick, green, 1.8–2.5 × 0.3–0.4 cm. *Inflorescence* axillary, 2–4 flowered (D); peduncle green, 1–1.5 × 0.4–0.5 cm; pedicel stout; bract swollen, smooth, light green (D). *Mature flower bud*: yellowish green, ovate, 1.4–1.6 × 0.8–1.1 cm, length to width ratio ca. 1.68 (not greater than 1.8), cross section four sided (F), apex obtuse (E); calyx lobes 4, thick, apex acute; petals 4, creamy white, thin, slightly hairy in the margin (I), linear, 1–1.2 × 0.3–0.4 cm; stamens 12–16, pale brown, occur in two distinct whorls (J), outer 8 long stamen, 1cm long, inner smaller stamens (G) varied from 2–8, 0.4–0.6 cm long; style bilobed, 0.1–0.15 cm long, seated on domed ovary (H); *Mature fruits*: rarely produced, only one was collected, 4 cm long, 2.3cm wide, dark brown (N); *Mature hypocotyls*: 29 cm long, green smooth, tip bluntly pointed, 1.5 cm wide at widest point (L); plumule green, 1.5 cm long, similar to hypocotyls of *R. apiculata* (M).

Morphological features of *Rhizophora* × *annamalayana*. (A) Habit; (b) leafy rosette with flowers; (C) broadly elliptic leaf; (D) mature bud; (E) closed mature flower bud; (F) mature bud cross section; (G) inner smaller stamens; (H) style; (I) mature propagule; (J) glabrous petal; (K) stamens in two whorls; (L) bark; (M) comparison of propagules *R.* × *annamalayana*, *R. apiculata*, *R. stylosa*, *R. mucronata* (left to right); (N) comparison of fruits *R.* × *annamalayana*, *R. apiculata*, *R. stylosa*, *R. mucronata* (left to right); (O) acute leaf tip with pointed mucronate.

Distribution

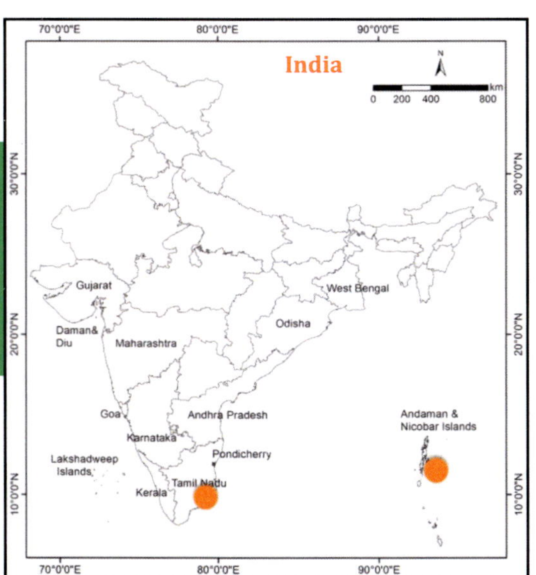

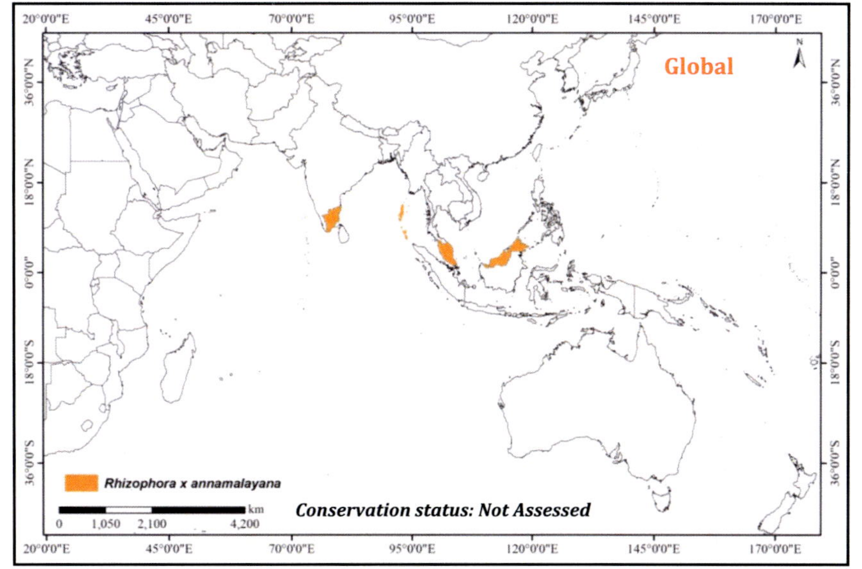

Habitat and Ecology

It is commonly found at mid intertidal areas and intermediate and downstream estuarine position in association with *R. apiculata* and *R. mucronata.*

Phenology

Flowering occurs throughout the year. Fruits are rarely present.

Notes

This is named after Annamalai University for its service for educating rural masses. Its identity and parentage are confirmed using molecular markers. Both *R. apiculata* and *R. mucronata* seemed to be able to mother the hybrids (Parani et al. 1997; Lakshmi et al. 2002; Lo 2010; Sahu et al. 2015).

Rhizophora mucronata var. alokii P. Ragavan, PhytoKeys 52: 95–103 (2015).

Rhizophora mucronata var. *alokii* (Rhizophoraceae), a new variety of *Rhizophora* from the A & N Islands, India, has been described by Ragavan et al. (2015c) from Austin Creek, North Andaman. The new variety is remarkable in having four stamens (all other *Rhizophora* species have 8 or more than 8 stamens), laterally folded leaves, a short peduncle, thick leathery petals, and a four-sided ovary with a sessile style.

Velvety and hairy petal

KEY CHARACTERS

- Tree columnar to spreading, grows upto 15–20 m, evergreen, extensive stilt root
- Bark dark brown, rough, friable, fissured horizontally
- Leaves green to dark green, broadly elliptical, plenty of dark spots underside
- Inflorescences axillary, 2–6 flowered, bract and bracteoles minute, peduncle short
- Flowers ellipsoidal, creamy white, bract and bracteoles minute
- Calyx lobes 4, thicker than *R. mucronata*, yellowish white, apex acute; petals 4, thick, leathery, velvety and hairy on the margin, stamens 4, style < 0.15 cm, bi-lobed, seated on four-sided domed ovary
- Mature hypocotyls 40–60 cm long, green, tip pointed

Fruit with emerging propagule

Leaves

Bark

Inflorescences

Stamens

Rhizophora mucronata var. alokii P. Ragavan

Taxonomy

Tree: columnar to spreading, height to 20 m, evergreen (A). *Bark*: dark brown, friable, fissured horizontally (C). *Roots*: both stilt roots and aerial roots growing from lower branches, stilt roots are highly conspicuous arching above ground to 2 m (B). *Leaves*: simple, opposite, green to dark green, elliptical to broadly elliptical (D, E), laterally folded, underside with numerous dark spots, 7–13 × 4–8.5 cm, length to width ratio averaging 1.69 (not greater than 1.8), apex obtuse with pointed mucro, 0.4–0.5 cm long (F), base cuneate, margin entire; petiole green, 1.5–3 × 0.3–0.5 cm. *Inflorescences*: axillary, 2–6 flowered (G); bract and bracteoles minute (H); peduncle 2–3.5 × 0.3–0.5 cm; pedicel stout; *Mature flower*: ellipsoidal, creamy white (Fig. 46I), 1.4–1.6 × 0.7–0.9 cm, length to width ratio *ca.* 1.87, cross section slightly four-sided (J); calyx lobes 4, thicker than *R. mucronata*, yellowish white, apex acute; petals 4, thick, leathery, folded laterally, creamy white, velvety and hairy on the margin (L, N), 0.9–1.1 × 0.3–0.4 cm; stamens 4, 0.5–0.7cm long (Fig. 46M, N, R); style bilobed (O), 0.8–0.12 cm long, seated on four sided domed ovary (P). *Mature fruits*: pear-shaped, brown, 4–5 × 2.5–3.5 cm, calyx persistent with erect lobes (Q). *Mature hypocotyls*: 40–60 cm long, green, tip pointed, 1.5–1.7 cm wide at widest point (K); plumule green, 2–3 cm long.

Distribution

Found only in Andaman Islands

Notes

This is a variety of *R. mucronata,* named in honour of Dr. Alok Saxena (Rtd. Principal Chief Conservator of Forests, A & N Administration) for his contribution to mangrove conservation in the A & N Islands. This taxon is distinct from other *Rhizophora* species in stamen number, but molecular evidence is yet to be provided (Ragavan et al. 2015c).

Distribution in India

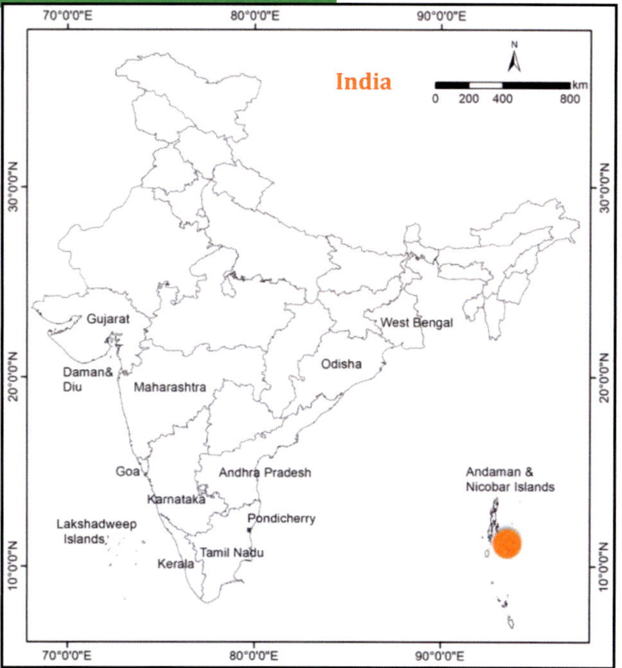

Habitat and Ecology

It grows in a mangrove forest along the banks in an intermediate estuarine position in association with *Rhizophora apiculata*, *R. mucronata* and *Ceriops tagal*.

Phenology

Flowering December to March; Fruiting April to July.

Morphological features of *Rhizophora mucronata* var. *alokii*. (A) Habit; (B) stem base with stilt root; (C) bark; (D) branches; (E) leafy rosette with flowers; (F) mucronate; (G) inflorescences; (H) minute bract at dichotomous branch; (I) mature bud with minute bracteole below calyx; (J) four sided cross section of bud; (K) mature propagules; (L) thick leathery petal; (M) stamens; (N) flower; (O) four sided ovary; (P) style; (Q) pear shaped fruit; (R) stamens with pollen..

Rhizophora ×mohanii P. Ragavan, ISME electronic Journal 13: 3-7 (2015) — Rhizophoraceae

Rhizophora × mohanii, a putative hybrid between *R. stylosa* and *R. mucronata*, described by Ragavan et al. (2015e) based on intermediate morphological characters from Andaman Islands, India. The hybrid *R. × mohanii* resembles *R. mucronata* by its broad leaves, and resembles *R. stylosa* by its long style (0.3–0.5 cm), but differs from *R. stylosa* and *R. mucronata* by its ovate leaves with rounded leaf apex, leathery texture, conspicuous bracts and bracteoles with one or two lobes, large dimension of mature buds (1.7 × 0.7 cm) and lack of propagules.

Long style with obscure ovary

Habit

KEY CHARACTERS

- Tree columnar to spreading, up to 10 m tall, evergreen, extensive stilt root
- Bark dark brown to grey, rough, friable, fissured horizontally
- Leaves dark green, leathery, oblong rounded, clustered at the end of branches, apex rounded with pointed mucronate tip.
- Inflorescences axillary, 2–8 flowered with prominent bract and bracteoles
- Mature flower buds ellipsoidal, calyx 4 lobed, petals 4, white, lanceolate, hairy, folded, stamens 8, style bi-lobed, 0.4–0.5 cm long, ovary not prominent.
- Fruits and propagules not observed.

Hairy petal

Leaf with rounded apex

Thick leathery leaves

Bark

Inflorescences

Bract single lobe

Bract bi-lobed

Taxonomy

Tree: columnar to spreading, up to 10 m tall, evergreen (A). *Bark*: dark brown to grey, rough, friable, fissured horizontally (B). *Roots:* both stilt roots and aerial roots growing from lower branches, stilt roots are highly conspicuous arching above ground to 2 m. *Leaves*: simple, opposite (E), oblong rounded (C), dark green and leathery, 0.1–0.15 cm thick, 8–14 × 6–9 cm, ratio of length to width 1.36, apex rounded (F) with pointed mucronate, to 0.3–0.4 cm long (G), base rounded, margin entire, laterally folded (D), leaves clustered at the end of branch; petiole green, 1.5–3.0 × 0.4–0.6 cm. *Inflorescences:* axillary, 2–8 flowered (H); bract prominent, either two lobed or single lobed (J); bracteoles prominent, two lobed (K); peduncle green, 2–5 × 0.3–0.4 cm, single lobed bract present before the first dichotomous branch (I); pedicel stout. *Mature flower buds:* yellowish green, ellipsoidal (K), four sided in cross section (L), 1.5–1.8 × 0.6–0.9 cm, ratio of length to width 2.21, apex obtuse, widest near the base; calyx lobes 4, yellowish white, apex acute; petals 4, white, lanceolate, hairy, folded, 1.2–1.3 × 0.3–0.4 cm (M); stamens 8, pale brown, 1.0 cm long (O); style bi-lobed, 0.4–0.5 cm long (N); ovary not prominent. *Fruits:* not observed.

Distribution in India

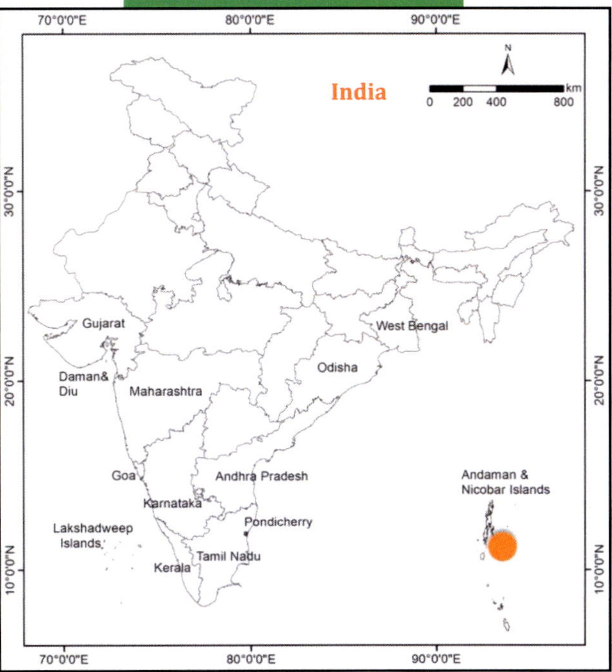

Habitat and Ecology

Found at low intertidal position along with *R. mucronata* and *R. stylosa.*

Phenology

Flowering occurs throughout the year. No other reproductive material was observed.

Distribution

Found only in Andaman Islands

Rhizophora × mohanii P. Ragavan

Morphological features of *Rhizophora × mohanii.* (A) Habit; (B) bark; (C) leathery ovate leaf; (D) laterally folded leaf; (E) leafy rosette with flowers; (F) rounded leaf apex; (G) mucronate tip; (H) inflorescences; (I) single bract like structure on peduncle before dichotomous branch; (J) dichotomous branch with prominent bract; (K) Mature bud with prominent bracteoles; (L) four sided cross section of mature bud; (M) flower with hairy petal; (N) style; (O) stamens.

Notes

This is named in honour of Prof. P.M. Mohan (Department of Ocean studies and Marine Biology, Pondicherry University, Port Blair, Andaman and Nicobar Islands). This taxon is a sterile hybrid based on the morphological intermediate characteristics of *R. mucronata* and *R. stylosa* without fruit settings. Molecular evidence is yet to be provided to confirm this hybrid.

Rhizophora stylosa var. *andamanica*. P. Ragavan, 2015 | Rhizophoraceae

Species feature

Rhizophora stylosa var. *andamanica* is another new entity described from A & N Islands by Ragavan (2015). This taxon is distinct from other *Rhizophora* species by its inflorescences structure. It has 2-4 flowered inflorescences, pedicel longer than peduncle with 3-4 cup shaped bracteoles and with a kind of inflorescence which is quite different from that of *Rhizophora* species. Other distinguishing characters are laterally folded elliptic leaves, intermediate style length (0.2- 0.25 cm), and large dimension of mature bud and occurrence of a few mature propagules.

Pedicle with multiple bracteoles

KEY CHARACTERS

- Tree grows upto15 m height, evergreen, stilt root extensively developed
- Bark dark brown to black, rough, fissure horizontally
- Leaves elliptic, green, apex acute with pointed mucronate tip, folded laterally
- Inflorescences axillary cymose, 2-4 flowered, bracts and bracteoles prominent, pedicel longer than peduncle, 3–3.5 cm long, with 3–4 prominent bracteoles
- Mature flower bud ellipsoidal, apex obtuse; calyx lobes 4, petals 4, thin, white, lanceolate, hairy, folded at margin, stamens 8, style bi-lobed, seated on domed ovary, 0.2–0.3 cm long
- Mature fruits pear shaped, brown, mature hypocotyls green, tip pointed, 40–60 cm long

Cub shaped bract

Laterally folded leaves

Bark

Bud with prominent bracteoles

Petals

Bi-lobed style

Rhizophora stylosa var. *andamanica*

Taxonomy

Tree: columnar to spreading, height to 15 m, evergreen (A). *Bark*: dark brown to black, rough, fissure horizontally (O). *Roots*: both stilt roots and aerial roots growing from lower branches, stilt roots are highly conspicuous arching above ground to 2 m. *Leaves*: simple, opposite, elliptic, green, clumped at end of the branch (B), folded laterally (C), 10–13.5 × 6–8 cm, length to width ratio averaging 1.7 (not greater than 1.8), apex acute with pointed mucronate (D), 0.5–0.7 cm, base cuneate, margin entire; petiole green 2.5–4 × 0.3–0.4 cm. *Inflorescences:* axillary cymose, 2-4 flowered (E); bracts prominent (N); bracteoles prominent cub like just below each flower (H); peduncle 2–3 × 0.3–0.4 cm; pedicel longer than petiole, 3–3.5 cm long, with 3–4 prominent bracteoles (E, F, G). *Mature flower bud*: yellowish green, ellipsoidal (H), 1.6–1.9 × 0.7–0.9 cm, ratio of length to width ca. 2.29 (not less than 2), apex obtuse, widest near the base; calyx lobes 4, greenish white, apex acute, cross section slightly four sided (K); petals 4, thin, white, lanceolate, hairy, folded at margin (I), enclosing stamens, 1–1.3 × 0.3–0.4 cm; stamens 8, pale brown 0.8–1 cm long (P); style bilobed, seated on domed ovary, 0.2-0.3 cm long (L). *Mature fruits*: pear shaped, brown, 3–3.8 × 2–3 cm (M). *Mature hypocotyls*: green, tip pointed, 40–60 cm long (J), 1.5 cm wide at widest point; plumule, green, 2–3 cm long.

Distribution in India

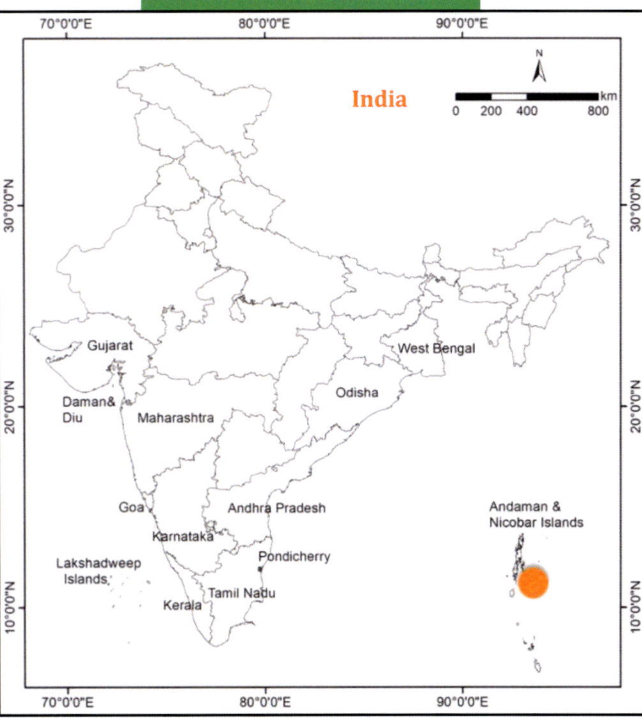

Habitat and Ecology

Found at low and mid intertidal position along with *R. mucronata* and *R. stylosa.*

Phenology

Flowering and Fruiting occurs throughout the year.

Distribution

Found only in Andaman Islands

Notes

This species is named after the type locality. This taxon is suspected to be an advanced stage of hybridization between *R. mucronata* and *R. stylosa*. Since *Rhizophora* species has wide phenotypic plasticity, molecular analysis is required to confirm the status of this taxon.

Morphological features of *Rhizophora stylosa* var. *andamanica*. (A) Habit; (B) laterally folded leaves; (C) leafy rosette with flowers; (D) mucronate; (E) inflorescences; (F & G) long pedicel with three bract like structures; (H) bud with prominent bracteoles; (I) flower with hairy petal; (J) mature propagules; (K) mature bud cross section; (L) style; (M) pear shaped fruit; (N) prominent bract at dichotomous branch; (O) bark; (P) stamens.

Scyphiphora hydrophylacea C.F.Gaertn., Suppl. Carp. 91 (1806). Rubiaceae

Scyphiphora hydrophylacea belongs to the family Rubiaceae. It is most common in A & N Islands, and it has restricted occurrence in West Bengal, Odisha and Andhra Pradesh. It is an erect, evergreen, often much-branched shrub or small tree. In the field *S. hydrophylacea* is easily recognized by its rounded glossy leaves, bunch of small white flowers arising from the leaf axial and drupe like fruits. Terminal nodes and shoots are also distinctively covered by a resinous substance. Occurs on mud, sand and rocky substrates on the landward margin of mangroves or on the banks of tidal waterways.

Branchlet with punch of axial flowers and fruits

KEY CHARACTERS

- Erect Shrub or small tree, grows up to 3m tall, occasionally stilt root present
- Leaves glossy green broadly drop- shaped, young leaves coated in a varnish like substance.
- Inflorescences axial, multiple flowered, peduncle short
- Flowers occur in dense clusters, tiny, petal 4, white, style exceeds the petals, bilobed
- Fruits broadly elliptic, deeply ridged lengthwise, light brown when ripe, buoyant.

Varnish coated like leaves

Inflorescences

Flower with bi-lobed style

Species feature

Pale green and cylindrical fruits

Scyphiphora hydrophylacea C.F.Gaertn

Taxonomy

Shrub: spreading, height to 2.5 m, multi-stemmed, evergreen (A). *Twigs*: green become brown on maturity. *Bark*: light brown, fissured (C). *Roots*: occasionally above ground, net like cable roots (I), stem base simple. *Leaves*: simple, opposite, margin entire, yellowish green (C), polished, leathery, obovate, 3.5–6.5 × 2–4 cm, ratio of length to width ca. 1.7, apex rounded slightly emarginate, base attenuate; petiole yellowish green, 1–2 cm long. *Inflorescence*: axillary, clustered (B), 5–10 flowers. *Mature bud*: ellipsoidal (F), 1 cm long; peduncle 0.3–0.5 cm; calyx tube green, 0.3–0.5 cm long; corolla tube 0.4-0.6 cm long, white or slightly pink petal lobes 4 (G), bluntly pointed, 0.3 cm lobes, reflexed at anthesis; stamens 4, at the mouth of corolla tube (H), 0.2–0.4 cm long; style slender with bilobed stigma (G, H), 0.5–0.8 cm long. *Mature fruits*: drupe shaped, four seeded, green turning brown on maturity with longitudinal ridges (E, J), 0.5–1 cm long.

Distribution

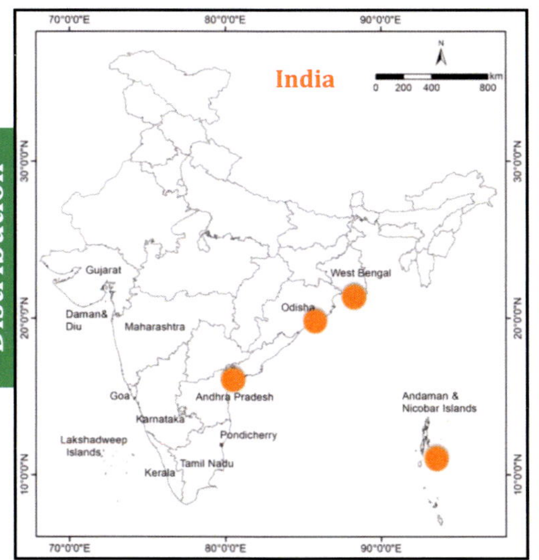

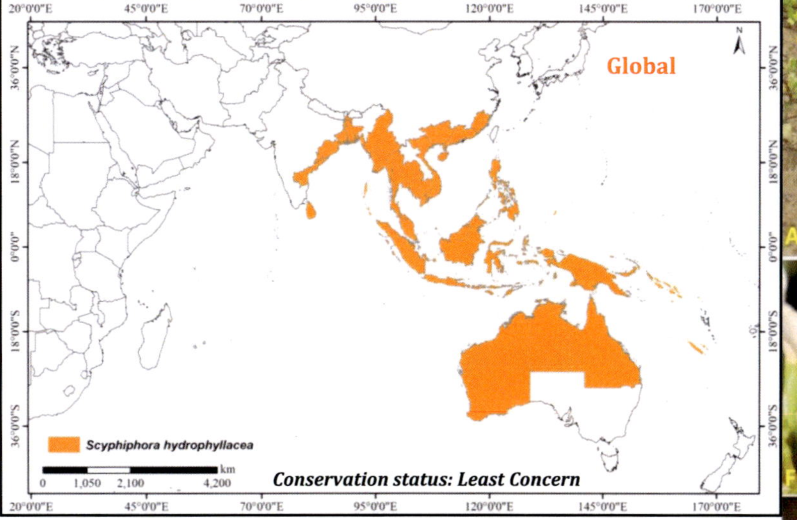

Habitat and Ecology

Often found in high intertidal and intermediate estuarine position on mud, sand and rocky substrates on the landward margin of mangroves or on the banks of tidal waterways.

Phenology

Flowering and fruiting occur throughout the year.

Notes

The identity of *Scyphiphora hydrophylacea* is often confused with *Scaevola taccada*, a shrubby coastal/beach vegetation, due to the presences of similar glossy green, resin coated glittering leaves. However, *S. hydrophylacea* has smaller sized leaves instead of larger ones as in *Scaevola taccada*.

Morphological features of *Scyphiphora hydrophylacea*. (A) Habit; (B) leafy rosette with flowers; (C) bark; (D) polished obovate leaves; (E) mature fruits; (F) mature bud; (G) cluster of flowers at leaf axial; (H) flower with reflexed petals, stamens and style; (I) above ground roots; (J) fruits.

Sonneratia alba

Genus: *Sonneratia*

Lythraceae

Genus *Sonneratia*, belonging to the family Lythraceae, is an important component of mangroves in the IWP region. It comprises of six species viz., *Sonneratia alba*, *S. apetala*, *S. caseolaris*, *S. griffithii*, *S. ovata*, and *S. lanceolata* and three natural hybrids *S.* ×*urama*, *S.* ×*gulngai* and *S.* ×*hainanensis* (Duke 2006; Tomlinson 2016). In addition, unnamed putative hybrids between *S. alba* and *S. griffithii* and *S. alba* and *S. apetala* are also known. Except *S.* ×*hainanensis*, all the other *Sonneratia* species are known from Indian mangroves (Ragavan et al. 2016).

Key to *Sonneratia species*

1. (a) Petals present..........3
   (b) Petals absent............4
2. (a) Petals white, stamens white......***S. alba***
   (b) Petals red, stamens red or white......... 3
3. (a) Stamens red, leaves elliptic or oblong with pointed mucronate tip.......***S. caseolaris***
   (b) Stamens red, leaves obovate to ovate with recurved mucronate tip........***S. ×gulngai***
   (c) Stamens red, leaves lanceolate, apex acute with recurved mucronate tip...***S. lanceolata***
   (d) Stamens white, leaves elliptic to broadly elliptic, apex acute with recurved mucronate tip...***S. ×urama***
4. (a) Leaves lanceolate, calyx four lobed................................***S. apetala***
   (b) Leaves broadly ovate, mucronate tip absent......................***S. ovata***
   (b) Leaves obovate to sub-orbicular, base rounded, mucronate tip present folded underside of the leaf..***S. griffithii***

Conical pneumatophores

Flower of *S. alba*

Flower of *S. caseolaris*

Flower of *S.* ×*urama*

Flower of *S. ovata*

Flower of *S. griffithii*

Sonneratia alba Sm., Cyclopedia 33 (1816).

Sonneratia alba is most common in A & N Islands, whereas it has restricted distribution along the East and West Coasts of Mainland India. It is often found in downstream of the tidal creek or low intertidal areas in association *Avicenna marina*, *Rhizophora* spp., and *Aegiceras corniculatum*. Exhibits wide morphological variation as well as coloration in floral parts.

Fissured brown color bark

KEY CHARACTERS

- Tree grows up to 10–15m height, prominent, conical pneumatophores present
- Bark is cream coloured to brown, with smooth, fine, longitudinal fissures
- Leaves opposite, blade dull light green, broadly drop-shaped.
- Inflorescences 1–3 flowered, terminal or axial
- Flower buds ellipsoidal with medial constriction,
- Calyx is cup shaped, inner side often red tinged, petals and stamens are white
- Fruits somewhat pear- shaped with persistent calyx, calyx lobes bend back towards the stalk, seeds are sickle shaped.

Underside curved mucronate

Drop shaped leaves

Species feature

Flowers with numerous white stamens

Mature fruits with reflexed calyx lobes

Flower with white petal and calyx lobes tinged red inner side

Sonneratia alba Sm.

Taxonomy

Tree: columnar or spreading, height to 10–15 m, evergreen (A). *Bark*: smooth grey in young (E), pale brown, fissured and flaky on maturity (D). *Roots*: pneumatophores (F), cone shaped, up to 40 cm long, tip bluntly pointed, branched, base stout, stem base simple. *Leaves*: simple, opposite, leathery, pale green, dull upper, satiny below, obovate-ovate (B), apex rounded with small thickened mucro recurved underside (C), base cuneate, 5–10 × 3–7 cm, margin entire; petiole 0.5–1.5 cm long, pale green to tinged red, terete. *Inflorescences*: terminal or axillary, 1–3 flowered. *Mature flower bud*: ellipsoidal, slightly constricted medially (G), apex acute to obtuse, 2–3.5 × 1.5–2 cm; calyx lobes 6–7, 1–2 by 0.5–1 cm, apex acute, inner often reddish (I, J); petals 6, rarely 5 or 7, white occasionally tinged green or red at base, linear or spathulate to stamen like (H, I), membranous if linear, 1–2 × 0.1–0.2 cm; stamens numerous, white, 2–5 cm long (H); ovary multi-locular; style terete, green, coiled in bud, extended at anthesis to 5 cm long, stigma fungiform (J), 0.3 cm wide. *Mature Fruit*: berry like, globose, 1.5–2.5 × 2.5–4.5 cm, persistent with style, pericarp green, globose, dull (L); calyx persistent on fruit, cup shaped, lobe spreading when young and reflexed towards the fruit stalk on maturity (K); seeds numerous, sickle shaped 0.8–1.2 cm long (M).

Morphological features of *Sonneratia alba*. (A) Habit; (B) leaf rosette with bud; (C) round apex of leaf with recurved mucronate; (D) mature fissured bark; (E) young smooth bark; (F) conical pneumatophores; (G) mature bud; (H) flower with white petal and stamens; (I) calyx lobes tinged red inside; (J) flower after anthesis; (K) mature fruit with reflexed calyx lobes; (L) cup shape calyx; (M) sickle shaped seeds.

Distribution

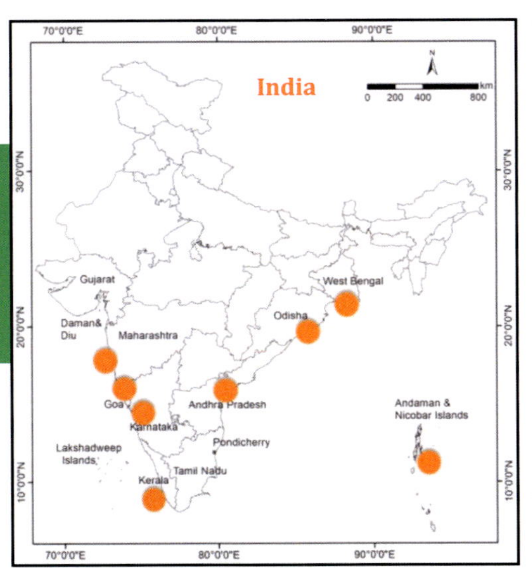

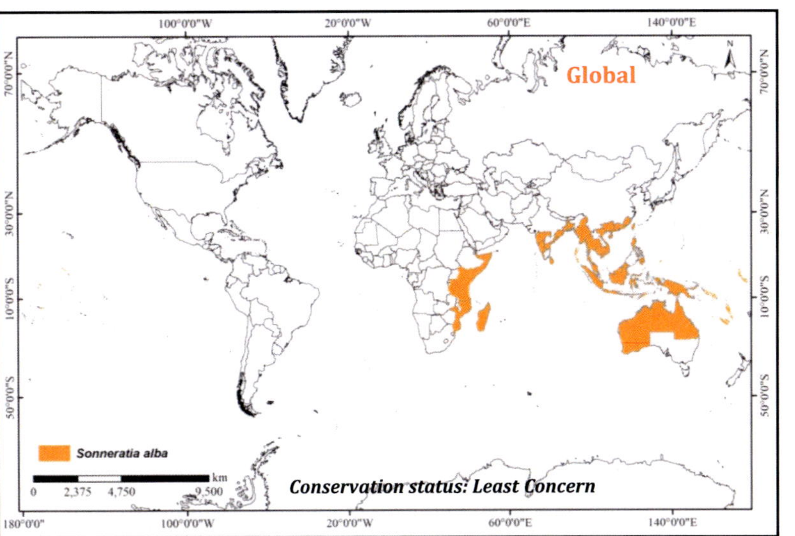

Habitat and Ecology

Often found in low intertidal and downstream estuarine on muddy substratum in association with *Avicennia marina*, and *Rhizophora* spp.,

Phenology

Flowering March to May; Fruiting June to September.

Notes

This taxon has significant morphological variations in bud shape, petal shape and petal colour. Petal colour is often white, but also pink or red tinged underside. Instead of linear thread like petal, ribbon like petals also occur. Inner side of calyx lobe is white or red tinged.

Sonneratia apetala Buch.-Ham., Account Embassy Kingd. Ava ed. 1, 477 (plate) (1800). Lythraceae

Sonneratia apetala is a widely distributed mangrove species in west and east coasts of Mainland India and not present in A & N Islands. It is often found in mid and upstream areas of the tidal creek or low and mid intertidal areas in association *Avicenna marina* and *A. alba*. In the field it easily distinguished from other *Sonneratia* species by its drooping branches, elongate and lanceolate leaves, small flower with four lobed calyx and umbrella shaped stigma.

Fissured bark

KEY CHARACTERS

- Tree grows up to 15–20m height, prominent, conical pneumatophores present
- Bark dark brown to black coloured, smooth when young, deeply fissure and peeling in thick flakes on maturity.
- Leaves elongate, lanceolate, blade dull light green, branchlets highly drooping
- Inflorescences, terminal or axial, often 3 flowered,
- Flower buds ellipsoidal with medial constriction,
- Calyx is cup shaped, 4–5 lobed, petals absent, stamens are creamy white, style long with prominent umbrella shaped stigma.
- Fruits small, berry like with persistent calyx

Pneumatophores

Lanceolate leaves

Inflorescences

Umbrella shaped stigma

Fruits

Sonneratia apetala Buch.-Ham.

Taxonomy

Tree: columnar or spreading, height to 15–20 m, evergreen (A). *Bark*: bark brown to black, young, fissured and flaky on maturity (G). *Roots*: pneumatophores, cone shaped up to 40 cm long(H), tip bluntly pointed, branched, base stout, stem base simple. *Leaves*: simple, opposite, leathery, pale green, dull upper, satiny below, elongate and lanceolate (B), apex acute, base cuneate, 5–10 × 2–5 cm, margin entire; petiole 0.5–1.5 cm long, pale green, terete. *Inflorescences*: terminal or axillary, 3–5 flowered (C). *Mature flower bud*: ellipsoidal, slightly constricted medially (D), apex acute to obtuse, 1–2. × 0.5–1.5 cm; calyx lobes 4–5, apex acute, inner surface often creamy white (E); petals absent, stamens numerous, creamy white, 2–3 cm long (E); ovary multi-locular; style terete, green, coiled in bud, extended at anthesis to 5 cm long, stigma umbrella shaped (F), 0.5 cm wide. *Mature Fruit*: berry like, globose, 1–2.5 × 2–3 cm, persistent with style, pericarp pale green, globose, dull (I); calyx persistent on fruit, cup shaped, lobe spreading; seeds numerous, irregular.

Distribution

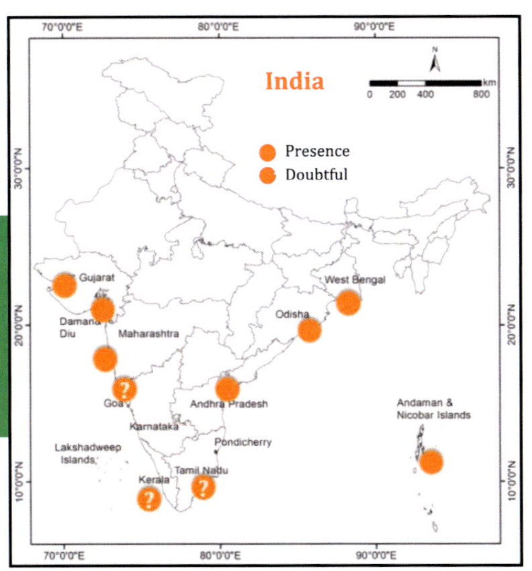

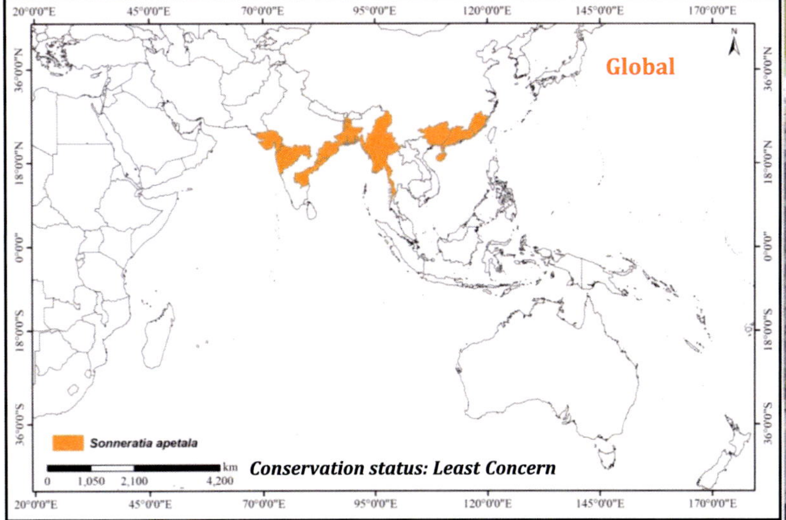

Habitat and Ecology

Often found in middle and upstream estuarine or mid and high intertidal region in association with *Avicennia marina*, *A. alba* and *Aegiceras coriculatum*.

Phenology

Flowering February to March; Fruiting April to May.

Notes

Occurrence of *S. apetala* in Goa, Kerala and Tamil Nadu needs to be verified. Natural hybrid is reported between *S. apetala* and *S. alba* from China (Xie et al. 2020). This putative hybrid is likely to be present along the presence of its parental species in Sundarbans, Odisha and Maharashtra. Reports of *S. apetala* from Andaman Islands are erroneous (Dagar et al. 1991; Debnath 2004).

Morphological features of *Sonneratia apetala*. (A) Habit; (B) lanceolate leaves (C) Inflorescences (D) mature flower bud; (E) opened flower with four lobed calyx, stamens and style (F) umbrella shaped stigma (G) matured bark (H) pneumatophores (I) fruits

Sonneratia caseolaris (L.) Engl., Pflanzenwelt Ost-Afr., C 286 (1895).

Lythraceae

Sonneratia caseolaris is common in east and west coasts of mainland India, whereas it has restricted distribution in A & N Islands. Often occur in upstream position of the tidal creek subjected to high levels of fresh-water runoff in association with *Barringtonia racemosa*. *Sonneratia caseolaris* is distinguished from other *Sonneratia* spp., by its red colour petals and stamens.

Flower with numerous red stamens and red petals

KEY CHARACTERS

- Tree growing in intertidal estuarine low saline areas
- Aerial roots are very prominent, thin slender conical pneumatophores, corky
- Bark is smooth with peeling of thin flakes at the base of the stem
- Leaves are simple, opposite, oval to oblong, short petiole with reddish pink base
- Mature flower bud ellipsoidal constricted medially, petals red, linear, membranous; stamens numerous red, rarely white in lower part.
- Fruit globose with persistent calyx and pointed style, calyx lobe flat, extending horizontally, not enclosing the fruit.

Oval shaped leaf

Multi-flowered inflorescences

Mature fruit with flat calyx

Mature bud with medial constriction

Sonneratia caseolaris (L.) Engl.

Taxonomy

Tree: columnar or spreading, height to 20 m, evergreen (A). *Bark*: smooth or finely fissured, grey to pale fleshy colour (H, M). *Roots*: pneumatophores (N), conical, thin, up to 50 cm long, tip narrow, flaky and branched, stem base simple. *Leaves*: simple, opposite, leathery, deep green (B), 4–12 × 2–6 cm, ratio of length to width <2, margin entire, apex acute to rounded with small thickened mucro recurved under (C, D); petiole up to 1 cm long, light green, tinged red towards base, flattened. *Inflorescence*: terminal or axillary, 1–3 flowered. *Mature flower bud*: ellipsoidal constricted medially (E), 2–2.5 × 1–1.5 cm, apex acute; calyx lobes 5–7, green, 1.5–2 cm long, apex acute, inner surface often red-streaked (I); petals 5–7, red, linear, membranous (F), 1.5–2 × 0.1–0.3 cm; stamens numerous, red or white or red in lower part , white in upper (G), 3.5 cm long; ovary multi-locular; style terete, green, coiled in bud, extended at anthesis to 6 cm long (I), stigma fungiform to 0.3 cm wide. *Mature Fruit*: berry broadly globose, to 3–4 × 2–5 cm, persistent with style (J); pericarp dark green, glabrous; calyx persistent, flat expanded (K); seeds numerous, angular irregular (O, P).

Distribution

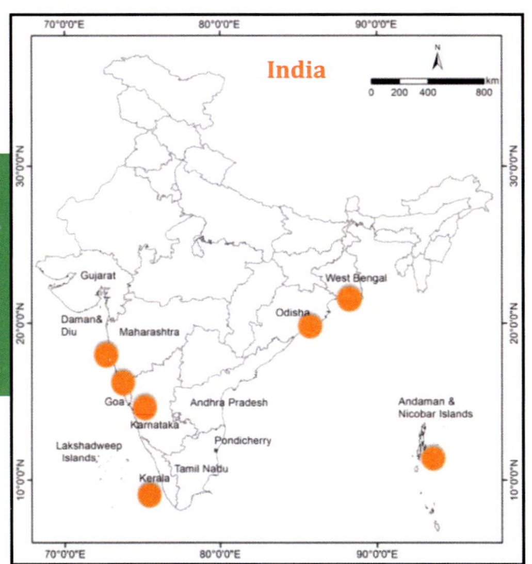

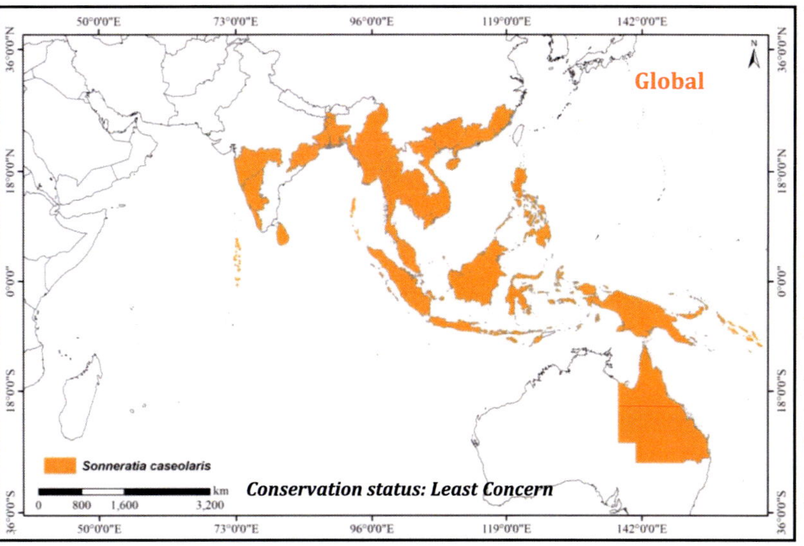

Habitat and Ecology

Found in low intertidal and upstream estuarine position. Often occurs in less saline parts of mangrove forests on deeply muddy soil, along tidal creeks with slow moving water.

Phenology

Flowering and fruiting occur throughout the year apparently.

Notes

Reports of *S. caseolaris* from Andaman Islands are erroneous (E.g. Dagar et al. 1991; Debnath 2004; Sreelakshmi et al. 2020a). This taxon exhibits highly variable leaf shapes and stamen colour (red or white or base red apex white). In A & N Islands, *S. caseolaris* is restricted to Little Andaman and Nicobar Islands.

Morphological features of *Sonneratia caseolaris*. (A) young seedling; (B) oblong leaves; (C) recurved mucronate; (D) leaf with rounded apex; (E) mature bud; (F) opened bud; (G) red petal and stamens; (H) smooth bark; (I) calyx lobes; (J) fruit with long style; (K) flat calyx; (L) stem base; (M) fine fissured bark; (N) thin conical pneumatophores; (O) cross section showing multi locular ovary with numerous seeds; (P) irregular seeds.

Sonneratia griffithii Kurz, Pegu For. Rep. App. B. 54, (1875); For. Fl. Brit. Burma, i. 527.

Sonneratia griffithii is one of the critically endangered species present in India. Its distribution is restricted only to Andaman Islands. It is easily distinguished from other species of *Sonneratia* by its obovate leaves, large solitary white flowers with white stamens, absence of petals and larger globose fruits with a depression at the apex and flattened calyx.

Mature flower with long style and numerous white stamens

KEY CHARACTERS

- Tree grows upto 20m height, prominent conical pneumatophore present
- Bark is deep or pale grey, fissured and flaky; smooth when young
- Leaves are obovate to almost round, mucronate tip recurved underside
- Intlorescences terminal, often 3-5 flowered, rarely solitary
- Mature flower bud rounded, apex blunt
- Calyx leathery, 6-7lobed, petals absent, stamens white
- Fruits are globose, calyx persistent and become woody, seeds numerous angular

Species feature

Ovate leaves

Mature rounded buds

Young fruit with flattened woody calyx

Mature fruit with flattened calyx

Sonneratia griffithii Kurz

Taxonomy

Tree: spreading, height to 25 m stem (A). *Bark*: pale brown, flaky, smooth when young and fissured on maturity (F). *Roots*: conical pneumatophores numerous (K), vertical, stout, elongate, cone-shaped, often branching, soft flaky surface, up to 60 cm long, stem base not buttressed. *Leaves*: simple, opposite, obovate to round, dark green (B), 7–11 × 6–10 cm, veins prominent on upper side, apex rounded with mucronate (C), base cuneate; petioles short, less 0.5 cm long. *Inflorescences*: solitary cymose on terminal and lateral branches, 3–5 flowers (D). *Mature flower buds*: globose, 2.5–3.0 cm long, with apex rounded (D); calyx tube 3.0–3.5 cm long and widely bell-shaped, calyx lobes 6–7 (H) do not envelop the base of the fruit, inner side of calyx is white (E); petals absent; stamens numerous, white, fall within hours after anthesis (G), anthers yellow, dorsi-fixed; ovary multi-locular. *Mature fruits*: berry like, globose with a depressed apex (I), 2.5–3.0 × 4.0–5.5 cm; pericarp leathery, style less persistent (L), calyx lobes spreading (J); seeds numerous and angular (M).

Distribution

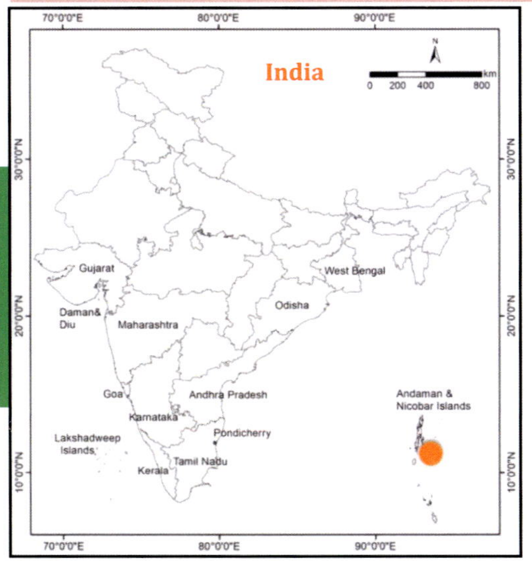

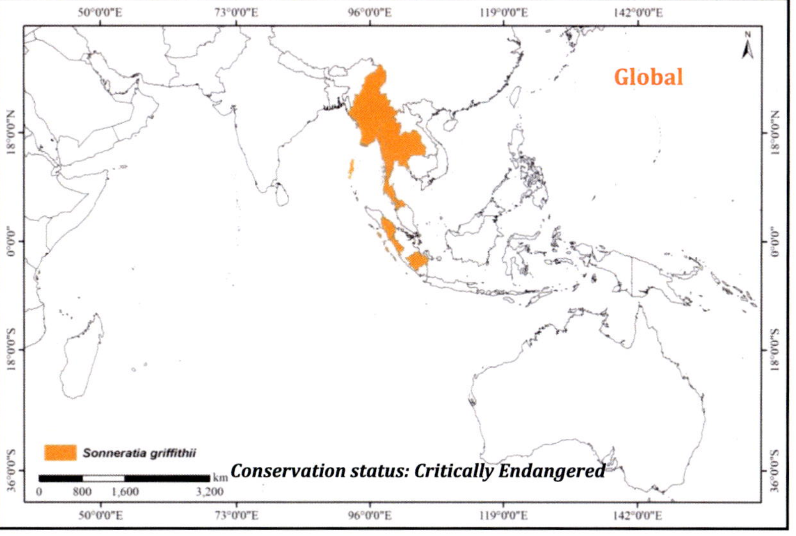

Habitat and Ecology

Ecology of this species is not fully known. This mangrove species is observed in high tide to mid tide locations in estuarine environment with soft mud substratum.

Phenology

Flowering February to April; Fruiting May to August.

Notes

Reports of *S. griffithii* from Sundarbans and Odisha are erroneous (e.g. Sreelakshmi et al. 2020b). It is restricted in distribution only to Andaman Islands in India. Specimens identified as *S. griffithii* from Odisha and Sundarbans are actually *S. alba* (Ragavan et al. 2019). Molecular evidence for the occurrences of putative hybrid between *S. alba* and *S. griffithii* has been known from Thailand.

Morphological features of *Sonneratia griffithii*. (A) Habit; (B) leaves; (C) recurved mucronate; (D) mature buds; (E) opened bud with white stamens; (F) simple stem base and fissured bark; (G) opened flower; (H) after anthesis; (I &J) mature fruit with flattened calyx; (K) pneumatophores; (L) young fruit; (M) fruit cross section.

Sonneratia ovata Backer, Bull. Jard. Bot. Buitenzorg ser. 3, ii. 329 (1920).

Lythraceae

Sonneratia ovata is a new addition to mangrove flora of India. First reported by Dam Roy et al. (2009) from Havelock Island. Subsequent floristic surveys revealed the occurrence of *S. ovata* in South, Middle, North Andaman Islands and Nicobar Islands (Ragavan et al. 2016). *S. ovata* is distinguished from other *Sonneratia* species by its fruits enveloped by calyx lobes.

Species feature

Fruits enveloped by calyx lobes

KEY CHARACTERS

- Tree grows upto 10m height, conical pneumatophores prominent
- Bark is pale brown to grey, fissured and slightly flaky, smooth when young
- Leaves are broadly ovate without mucronate tip
- Inflorescences are terminal either solitarily or in groups of three
- Flower bud oval shaped with medial constriction
- Calyx grooved along the lobes margin, inner side of calyx lobe often red tinged, petals often absent, if present it is white, inconspicuous, stamens white
- Fruits globose, calyx lobes enveloped the fruits, seeds are irregular.

Ovate leaf without mucronate tip

Inflorescences

Mature bud

White stamens

Young fruit with enveloped calyx

Sonneratia ovata Backer

Taxonomy

Tree: columnar with quadrangular young branches, height 10–20 m, evergreen (A). *Bark*: grey, fissured and flaky (B), smooth when young. *Roots*: conical pneumatophores (C), up to 40 cm long, stem base simple. *Leaves*: simple, opposite, green, ovate to broadly ovate (F), apex obtuse without mucronate tip, base rounded, 6–10 × 5–8 cm, ratio of length to width ca,1.2; petiole terete, reddened, 0.5–1 cm long, 0.4 cm. *Inflorescence*: terminal, 1–3 flowered (D). *Mature flower buds*: broadly oval and covered with small warts (J), 2.5–3 × 1.5–2 cm, apex obtuse; peduncle 1–1.5 cm long; calyx 6, 1.5–2 × 1 cm, leathery, verrucose outside, smooth, pinkish at base inside (E); petals absent, if present linear; stamens numerous, white, 3.5–4 cm long (G); style terete coiled in bud (J), extended at anthesis to 4–4.5 cm long (E), stigma fungiform; ovary multilocular. *Mature fruits*: berry like, globose (I), 2.5–4 × 3.5–5 cm; style and calyx persistent, calyx flattened, and lobes enveloped the fruit (H, K), style erect up to 5 cm long; seeds numerous, irregular.

Distribution

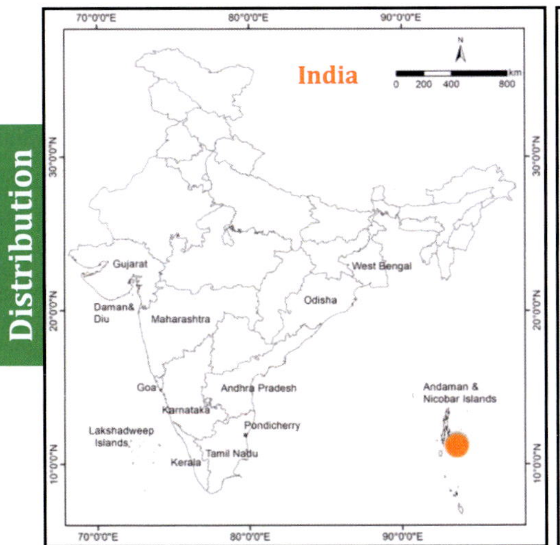

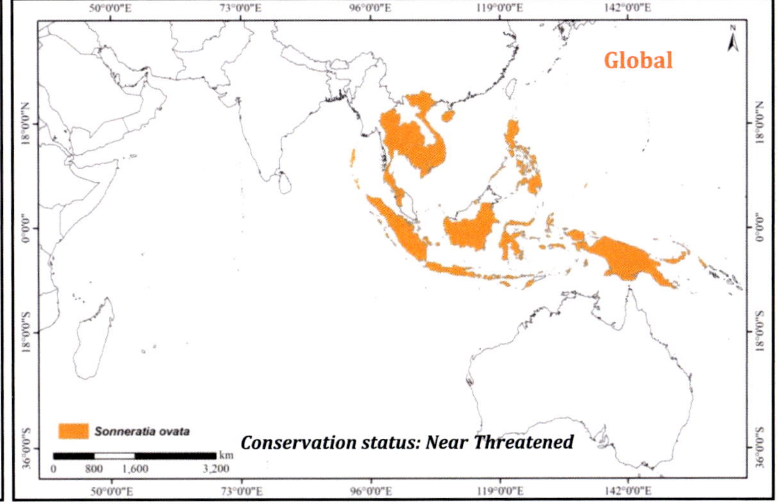

Habitat and Ecology

Often occurs on the landward edge or upstream estuarine position of mangrove swamps in brackish water and muddy soil.

Phenology

Flowering and Fruiting occurs throughout the year. Flowering peaks March to July; Fruiting August to December.

Notes

Sonneratia ovata is "Near Threatened" mangrove species. Without flowers and fruits, *S. ovata* looks like *S. griffithii.* However, leaves of *S. ovata* lacks mucronate tip, whereas *S. griffithii* have prominent mucronate tip.

Morphological features of *Sonneratia ovata*. (A) Habit; (B) fissured bark; (C) pneumatophores; (D) three flowered inflorescences; (E) calyx lobes tinged red inner side; (F) leaves; (G) opened flower; (H) young fruit; (I) mature fruit with enveloped calyx; (J) mature bud with white stamens; (K) enveloped calyx lobes.

Sonneratia lanceolata Blume, Mus. Bot. 1(22): 337 (1851).

Sonneratia lanceolata is a new record for India from Great Nicobar Island (Ragavan et al. 2014). *Sonneratia lanceolata* is distinguished from other *Sonneratia* species by its drooping branchlets, lanceolate leaves, ovoidal flower buds without medial constriction, white stamens, and red petals.

KEY CHARACTERS

- Tree grows upto 10–15m height, branches are highly drooping, conical pneumatophores present.
- Bark smooth or lightly fissured with numerous lenticels, grey to pale green
- Leaves simple, opposite, pale green, lanceolate with short red coloured petiole
- Inflorescence terminal, mostly 1 flowered rarely 2
- Flower buds are ovoidal or ellipsoidal, without medial constriction, apex acute to obtuse.
- Petal dark red, ribbon like; stamens red, numerous
- Mature fruits berry like with flattened calyx tube, calyx lobes erect.

Lanceolate like leaf

Smooth bark with lenticels

Pneumatophores

Flower bud

Red petal and stamens

Fruit with flattened calyx

Sonneratia lanceolata Blume

Taxonomy

Tree: Columnar or spreading, height to 25 m, branches are drooping (A, B), evergreen. *Bark*: smooth or lightly fissured with numerous lenticels, grey to pale green (F, G). *Roots*: pneumatophores (H), conical, slender up to 40 cm height, stem base simple (E). *Leaves*: simple, opposite, pale green, lanceolate with short petiole (C, D), 6–12.3 × 2.5–4.4 cm, ratio of length to width >2, leaf base attenuate, tip acute with recurved mucronate; petiole short 0.2–0.6 cm, flattened, red (C). *Inflorescence*: terminal, mostly 1 flower (I), rarely 2. *Mature flower buds*: ovoidal or ellipsoidal, without medial constriction, apex acute to obtuse (K, M), 2.2–3.5 × 1.5–2.4 cm; bract 2, lateral, not persistent (Fig. 33O); calyx lobes 5–7, apex acute (J); petal dark red, ribbon like (L), 1.8–2.2 × 0.25–0.3 cm; stamens red, numerous (N); ovary multilocular; style terete, coiled in bud (N), extended at anthesis to 5.1–5.6 cm long, stigma fungiform to 0.18–0.21 cm wide. *Mature fruits*: berry like (P), globose 2.1–2.5 × 4.5–5 cm, persistent withered style up to 6 cm long; pericarp green, leathery; calyx persistent flat expanded, calyx lobes erect (Q); seeds numerous angular irregular (R).

Distribution

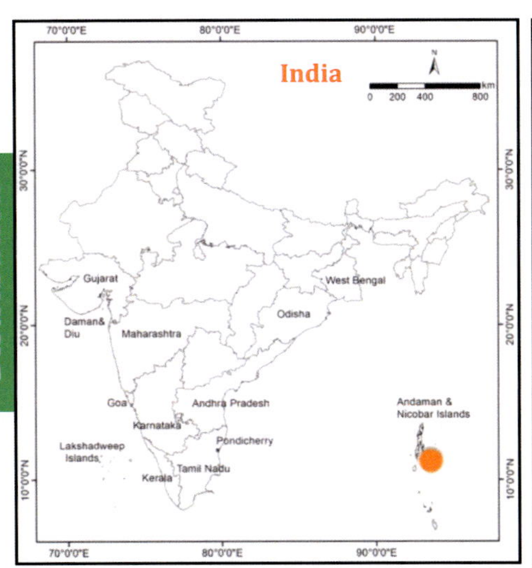

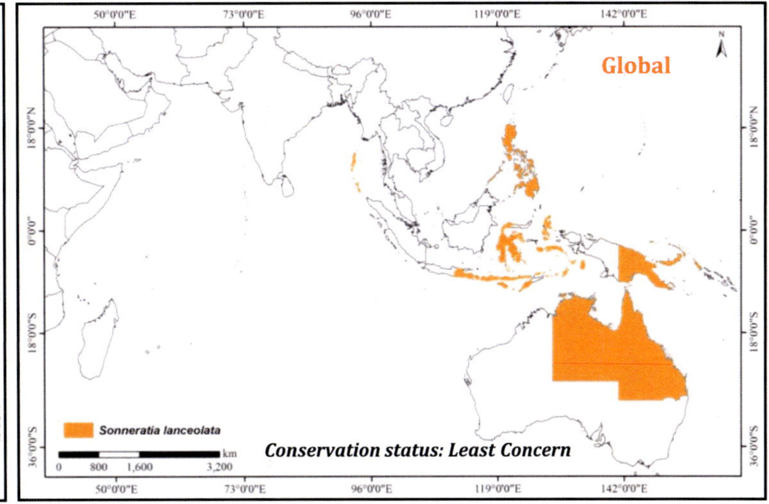

Habitat and Ecology

Found from low intertidal position on fine soft muddy substratum in association with *S. caseolaris.*

Phenology

Flowering and fruiting occur throughout the year.

Notes

Sonneratia lanceolata from Great Nicobar Island possess red stamen and red petals, whereas *S. lanceolata* described by Duke and Jackes (1987) and Duke (2006) from Australia has red petals and white stamens. Except the stamens colour all the other characters are resembling *S. lanceolata* as described by Duke and Jackes (1987).

Morphological features of *Sonneratia lanceolata*. (A) drooping branch; (B) branchlet with fruit; (C) lanceolate leaf with short petiole; (d) branchlet with terminal bud; (E) stem base; (F) Smooth grey bark; (G) lightly fissured bark; (H) pneumatophores; (I) opened bud; (J) dark red petals; (K) ovoidal bud without medial constriction; (L) red stamen and dark red petals of ovoidal bud; (M) ellipsoidal bud without medial constriction; (N) red stamen and Dark red petals of ellipsoidal bud; (O) mature bud with lateral bract; (P) Mature fruit; (Q) persistent flat expanded calyx; (R) angular irregular seeds.

Sonneratia ×urama N.C.Duke, Austral. Syst. Bot. 7(5): 522 (1994). Lythraceae

Sonneratia × urama is a putative hybrid between *S. alba* and *S. lanceolata* and known to occur in Indonesia, New guinea and Australia (Duke 2006). Ragavan et al. (2014) reported its occurrence in Great Nicobar Island and it represents new distribution record for India. *Sonneratia × urama* is distinguished from S. × *gulngai* and *S. caseolaris* by its white stamens and red petals whereas in S. × *gulngai* and *S. caseolaris* stamens and petals are red. Further it is distinguished from *S. lanceolata* by its prominent medial constriction in mature bud, elliptic leaf with broadly acute or obtuse apex and large sized fruits.

Fruit with flattened calyx

KEY CHARACTERS

- Large sized tree, mid intertidal or mid-estuarine position
- Bark smooth or lightly fissured with numerous lenticels, grey to pale green
- Leaves simple, opposite, pale green, elliptic to broadly elliptic
- Inflorescence terminal, 1 -2 flowers, mostly 1, flower bud ellipsoidal with prominent medial constriction, apex acute
- Petal dark red, ribbon like; stamens white, numerous
- Mature fruits berry like with flattened calyx tube

Elliptic leaf with recurved mucronate tip

Ellipsoidal bud

Flower with bract

Red petals and white stamens

Fruits with long style

Sonneratia ×urama N.C.Duke

Taxonomy

Tree: spreading with dense canopy, height to 20 m. Bark: smooth, grey, lightly fissured, flaky on maturity (G). *Roots*: pneumatophores, conical, slender up to 40 cm height, stem base simple (H). *Leaves*: simple, opposite, pale green, elliptic to broadly elliptic (A), 7–9.3 × 4.7–6.2 cm, ratio of length to width <2, leaf base attenuate, tip acute or broadly acute with recurved mucronate (B); petiole short 0.3–0.6 cm, flattened, red (C). *Inflorescence*: terminal 1–2 flowers, mostly 1 (C). *Mature flower bud*: ellipsoidal with prominent medial constriction (D), apex acute, 3.5–4 × 2–2.2 cm; bract 2, lateral, not persistent (D); calyx lobes 5–7, apex acute (E); petal 5–7, red, ribbon like, 2.3 × 0.2 cm (F); stamens white, 5.2 cm long, numerous (I, J); ovary multilocular; style terete, coiled in bud, extended at anthesis to 5.6–6.5 cm long (K); stigma fungiform to 0.19–0.21 cm wide. *Mature fruit*: berry like, globose 2–3.3 × 5.6–6 cm, persistent withered style up to 8.3 cm long, base convex (N); pericarp green, leathery; calyx persistent flat expanded, calyx lobes reflexed upwards (L, M); seeds numerous angular irregular.

Distribution

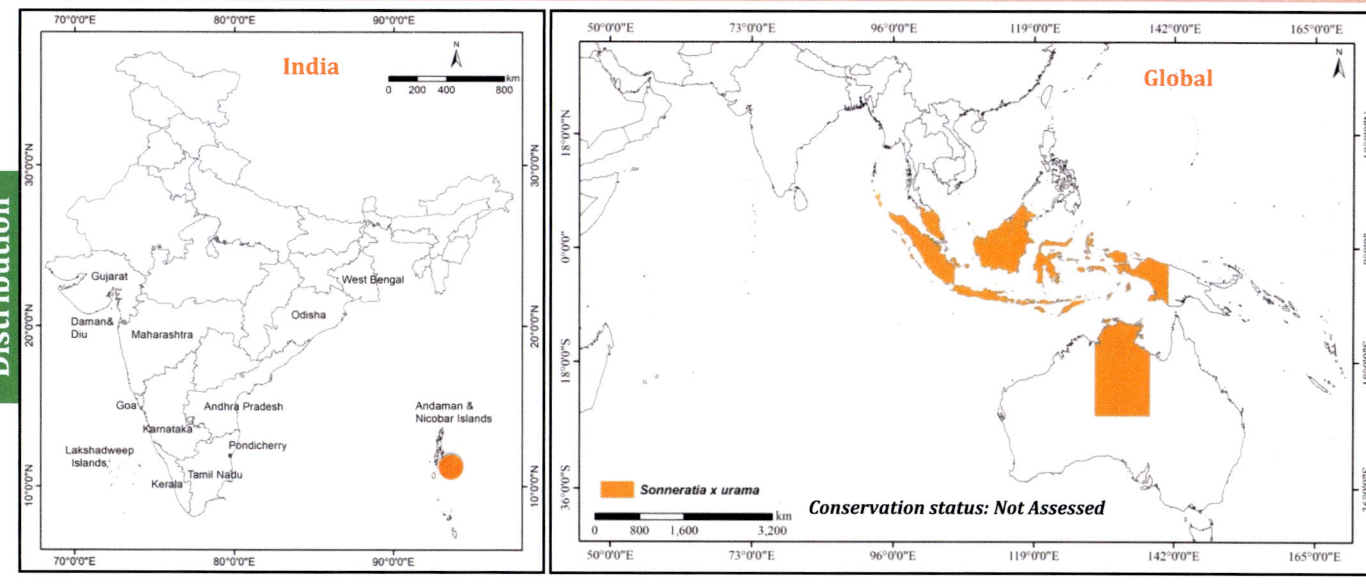

Habitat and Ecology

Found in downstream estuarine position and newly formed intertidal areas in Great Nicobar Island in association with *Sonneratia ×gulngai* and *Acrostichum aureum*.

Phenology

Flowering and fruiting occur throughout the year.

Notes

Since *S. caseolaris* also exhibits red petals and white stamens (rarely), identity of this taxon remains problematic. However, differences in habitat preferences between these two species will be highly useful to ensure the correct identity. *S. caseolaris* is often found in upstream low saline areas, whereas *S. × urama* is present in downstream high saline area.

Morphological features of *Sonneratia × urama*. (A) Branchlet; (B) elliptical leaf with acute tip; (C) flower bud with lateral bud; (D) Ellipsoidal bud with acute tip and constricted medially; (E) opened bud; (F) red petal and white stamens; (G) Smooth bark; (H) conical pneumatophores; (I & J) Flowers with long style; (K) cup shaped calyx in young fruit; (L) flat expanded calyx of mature fruit; (M) reflexed calyx lobes of Mature fruit; (N) fruits convex around the style base.

Sonneratia ×*gulngai* N.C.Duke, Austrobaileya 2(1): 103 (1984).

Branchlet with solitary flower

Sonneratia × *gulngai* is a putative hybrid between *S. alba* and *S. caseolaris*, known from Southern Malaysia and china through Southern Asia and Indonesia, to northern new Guinea and Northern Australia (Duke 2006). Recently Ragavan et al. (2014) reported its occurrence in Great Nicobar Island and it represents a new distribution record for India. *Sonneratia* × *gulngai* is distinguished from other *Sonneratia* spp., by its red stamens and petals, ovoid to obovate leaf with broadly acute tip, mature bud with medial constriction.

KEY CHARACTERS

- Tree grows up to 10–15m height, prominent conical pneumatophores present
- Bark smooth or lightly fissured with numerous lenticels, grey to pale green
- Leaves obovate to ovate, tip leaf tip broadly acute with recurved mucronate
- Inflorescence terminal, 1–2 flowers, flower bud ellipsoidal with prominent medial constriction, apex acute
- Petal dark red, ribbon like; stamens red, numerous
- Mature fruits berry like with flattened calyx tube and long style

Obovate leaf

Ellipsoidal bud with medial constriction

Red petals and stamens

Berry like fruit

Fruit with flattened calyx and long style

Sonneratia × gulngai N.C.Duke

Taxonomy

Tree: spreading with dense canopy, height to 20 m. *Bark*: smooth, lightly fissured and flaky on maturity, grey (I, J). *Roots*: pneumatophores (J), conical, slender up to 40-60 cm height, stem base simple. *Leaves*: simple, opposite, pale green, obovate to ovate (A), 7–9.3 × 4.7–6.2 cm, ratio of length to width <2, leaf base rounded or obtuse, leaf tip broadly acute with recurved mucronate; petiole short 0.2–0.5 cm, flattened, red. *Inflorescence*: terminal 1-2 flower (C), mostly 1. *Mature flower buds*: ellipsoidal with prominent medial constriction (D), apex acute, 3.8–4.4 × 2–2.3 cm; bract 2, lateral, not persistent; calyx lobes 5–7 (E), apex acute; petal 5–7, red, ribbon like (F), 2.5 × 0.2 cm; stamens red (G), 4.5–5.6 cm long, numerous; ovary multilocular, style terete, coiled in bud, extended at anthesis (H) up to 6.5 cm long; stigma fungiform to 0.17–0.19 cm. *Mature fruits*: berry like, globose 2.5–3.3 × 5.5–6 cm (B), style persistent, up to 8.6 cm long; pericarp green, leathery; calyx persistent flat expanded (K), calyx lobes reflexed upwards (L); seeds numerous angular irregular.

Distribution

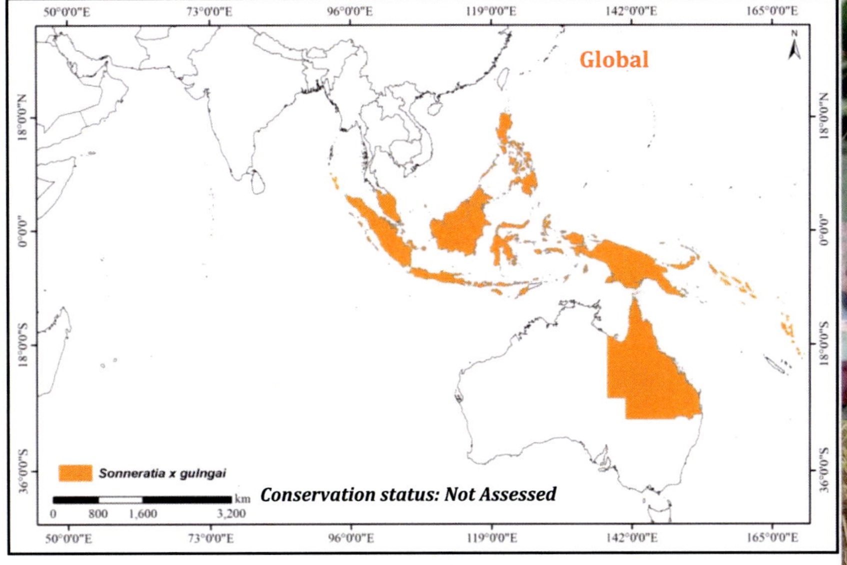

Habitat and Ecology

Found in downstream estuarine position and newly formed intertidal areas in Great Nicobar Island in association with *Sonneratia × urama* and *Acrostichum aureum.*

Phenology

Flowering and Fruiting occur throughout the year.

Notes

Since *S. caseolaris* also exhibits red petals and red stamens, identity of this taxon remains problematic. However, differences in habitat preferences between these two species will be highly useful to ensure the correct identity. *S. caseolaris* is often found in upstream low saline areas, whereas *S. × gulngai* is present in downstream or mid-estuarine position.

Morphological features of *Sonneratia × gulngai*. (A) Elliptic leaves with rounded tip; (B) branchlet with fruits; (C) inflorescences; (D) mature bud with acute tip and constricted medially; (E) red stamens and red petals; (F) ribbon like red petal; (G) opened flower; (H) flower after anthesis; (I) grey lightly fissured bark; (J) pneumatophores; (K) young fruit with cup shaped calyx; (left), mature fruit with flat calyx; (L) mature fruit with reflexed calyx lobes and convex style base.

Flower of *Xylocarpus granatum*

Genus: *Xylocarpus*

The genus *Xylocarpus* belonging to the Family Meliaceae has three distinct species, viz., *X. granatum*, *X. moluccensis* and *X. rumphii*. Of these, *X. granatum* and *X. moluccensis* are occurring in mangroves, while *X. rumphii* normally grows above high water on cliffs, rocks and sandy upland areas (Duke 2006). Recently Ragavan et al. (2015b) reported the occurrences of all three *Xylocarpus* species in the A & N Islands and highlighted the uncertainty in identity and nomenclatural ambiguity that exists between *X. moluccensis* and *X. rumphii.* It is difficult to distinguish *Xylocarpus* species based on herbarium specimens without observing the plant in situ. They can be easily recognized from each other in the field based on the characters of the root, trunk, bark, leaves, inflorescence and fruit.

Key for *Xylocarpus* species

1. (a) Above ground root present..........2

(b) Above ground root absent............3

2. (a) Leaflets 2-3 pairs, bark dark brown, fissured and peeling in narrow strips, fruits small, to the size of an orange, 7-10 cm in diameter, fruits with 8-16 seeds.......***Xylocarpus moluccensis***

b) Leaflets mostly 2 pairs, rarely 3 pair, bark smooth pale brown, peeling in thick flakes in patches, fruits large, 15-20 cm in diameter, fruits with 8-20 seeds.....***Xylocarpus granatum***

3. Leaf lets 2-4 pairs, bark fissured, not flaky, fruit about the size of an orange, 7-10 cm in diameter and fruits with 4-10 seeds....***Xylocarpus rumphii***

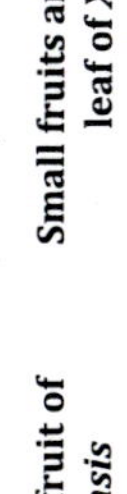

Small fruits and heart shaped leaf of *X. rumphii*

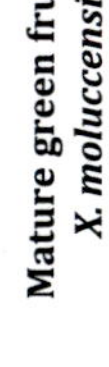

Mature green fruit of *X. moluccensis*

Large seized fruit of *X. granatum*

Xylocarpus granatum J.Koenig, Naturforscher (Halle) 20: 2 (1784).

Species feature

Xylocarpus granatum is most common in East coast of mainland India and A & N Islands. Generally, occurs in intermediate and upstream estuarine position along with *Rhizophora* spp., *Ceriops* spp., and *Bruguiera* spp. Easily recognized in the field by its buttressed plank root, smooth bark peeling in flakes or patches and large seized fruits. *Xylocarpus granatum* is often called as Sundari in A & N Islands, Whereas *Heritiera fomes* is called as Sundari in Sundarbans.

Thin flaky bark

KEY CHARACTERS

- Tree grows up to 10m height, stem base buttress and above ground planks roots present
- Bark is smooth, pale greenish or yellowish, peeling in thin flakes
- Leaves are compound, leaflet 2 pairs mostly, rarely 3 pairs, elliptical to obovate, leaf tip rounded, tapering at base
- Flowers are small, white, axillary, panicle up to 20 flowered, unisexual; calyx 4 lobed, yellow green, corolla -4 creamy to greenish white; stamens fused into tubes
- The disc around the ovary base is orange in colour
- Fruits are yellowish brown, ball shaped, 15 semi triangular seeds in a fruit with a corky testa.

Compound leaves with 2 pairs of leaflets

Inflorescences

Swollen ovary with orange colour disc

Mature fruits

Planks root

Taxonomy

Tree: spreading, height to 15 m, evergreen (A). *Bark*: smooth, flaky, thin, peeling in patches (H). *Roots*: above ground, ribbon-like plank roots; stem base buttresses (E). *Leaves*: compound, alternate (B), 10–17 cm long; petiole green, 2–4 cm long; leaflets oblong obovate (C), yellowish green, mostly 2 pairs rarely 3 pairs, 6–13 × 3.5–6.5 cm, ratio of length to width >2, leaflet stalk 0.5–1 cm long. *Inflorescence*: axillary, multiflowered panicle branched irregularly (F). *Mature flower bud*: rounded (G), 0.3–0.5 cm long; calyx lobes 4, green (G), minute, 0.3 cm long; petals 4, white, oval, 0.5–0.7 × 0.3 cm; *stamens* united and forming a staminal tube (G), white, 0.5 cm long, 8 lobes in upper margin, anther sessile yellowish-green present inside the staminal tube; style 0.3cm long, seated on raised ovary, stigma disc shaped; ovary superior, orange red in colour (D); in male flowers ovary is slender and non-functional, in female flowers stamens are non-functional. *Mature Fruits*: woody, 4 chambered, green turning brown on maturity (I, J), 15–20 cm in diameter; seeds 8–20, angular, smooth, tetrahedral, pale brown (K), up to 10 cm long.

Distribution

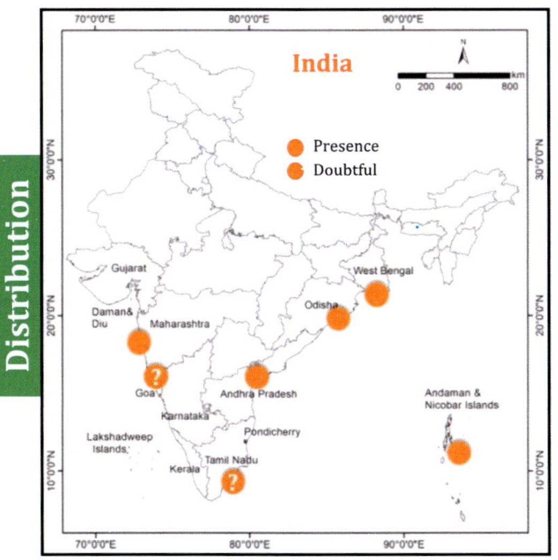

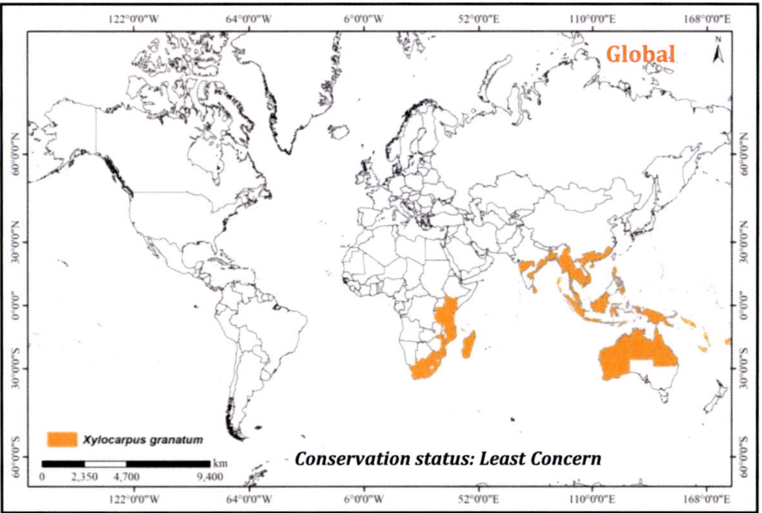

Habitat and Ecology

Often occurs in intermediate and upstream estuarine position, also occurs along banks of tidal creeks in association with *Rhizophora* spp., *Bruguiera* spp.

Phenology

Flowering February to March; Fruiting April to July.

Notes

Species of *Xylocarpus* are mentioned earlier under the generic name of *Carapa*. For instance, Parkinson (1923) described *Xylocarpus* species of A & N Islands under the generic name *Carapa*. He recognized three species, namely, *Carapa obovata*, *C. moluccensis* and *Carapa* sp., of which *Carapa obovata* is synonym to *X. granatum*.

Xylocarpus granatum J. Koenig

Morphological features of *Xylocarpus granatum*. (A) Habit; (B) leaf with two pairs of leaflet; (C) leaflet with rounded apex; (D) style seated on swollen ovary; (E) buttress; (F) inflorescences; (G) flowers with petals and fused stamens; (H) bark; (I) young fruit; (J) ripened fruit; (K) seeds.

Xylocarpus moluccensis (Lam.) M. Roem. Fam. Nat. Syn. Monogr. 1: 124 (1846). Meliaceae

Xylocarpus moluccensis is rare and known from East Coast of mainland India and Andaman Islands. In the field it is easily recognized by its rough bark with peeling in long narrow strips, prominent peg-like pneumatophores, 2-3 pairs dark green leaflets with acute tip, fruits green and smaller than *X. granatum*. Generally, occur in intermediate estuarine position.

Sub-globose fruits

KEY CHARACTERS

- Tree grows up to 20m in height and deciduous.
- It has prominent conical peg like pneumatophores
- Bark is light brown to grey and fissured with thick long narrow peelings
- Leaves are compound, 2–3 pairs of dark green leaflets with acute tip
- Inflorescence axillary with many flowers like *X. granatum*
- The disc around the ovary base is red in colour
- Fruits is sub-globose, green smaller than *X. granatum* with 8-16 pyramidal seeds

Species feature

Leaflets with acute tip

Conical peg like pneumatophores

Axial inflorescences

Flower with white petals and fused stamens

Ovary with red colour disc

Xylocarpus moluccensis (Lam.) M. Roem.

Taxonomy

Tree: columnar or spreading, height to 20 m (A), deciduous. *Bark*: brown to grey, fissured with thick flakes (J). *Root*: peg-like pneumatophores (I), apex blunt, erect 10–20 × 5 cm, stem base simple, occasionally small buttresses. *Leaves*: compound, alternate, 8-15 cm long; petiole 2–5 cm; leaflets 2–3 pairs (B), dark green, broadly elliptic, apex broadly acute to obtuse, base cuneate (C), 5–11 × 3–6 cm, ratio of length to width >2, leaflet stalk 0.5–1 cm. *Inflorescence*: axillary, multi-flowered panicle, monoecious, male and female flowers present in same inflorescence (F); peduncle 0.2–0.5 cm long. *Mature flower bud*: 0.3–0.5 cm long, tetramerous, white; calyx lobes 4, green, united below, 0.2 cm long; petals 4, creamy white, 0.3 × 0.1 cm, oblong; stamens united as staminal tube, creamy white, 0.3 cm long, upper margin bears 8 erect lobes (D); style 0.3 cm long, seated on raised ovary, stigma disc shaped; ovary superior, red in colour (E); in male flowers ovary is slender and non-functional, in female flowers stamens are non-functional. *Mature fruits*: woody, green, 4 chambered, green, 7–10 cm in diameter (G); seeds 8–16, angular, smooth, tetrahedral, pale brown, up to 10 cm long (H).

Distribution

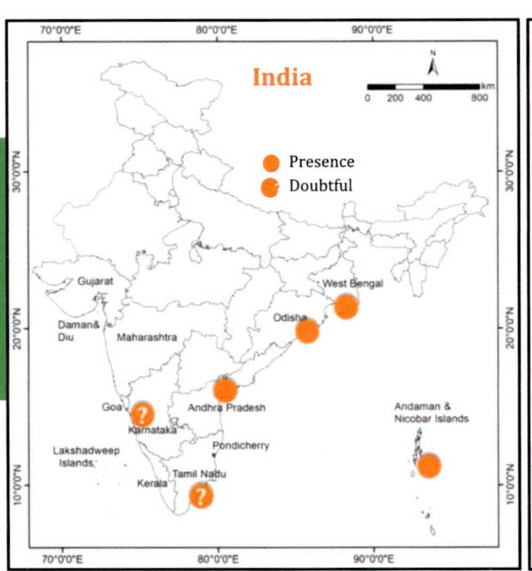

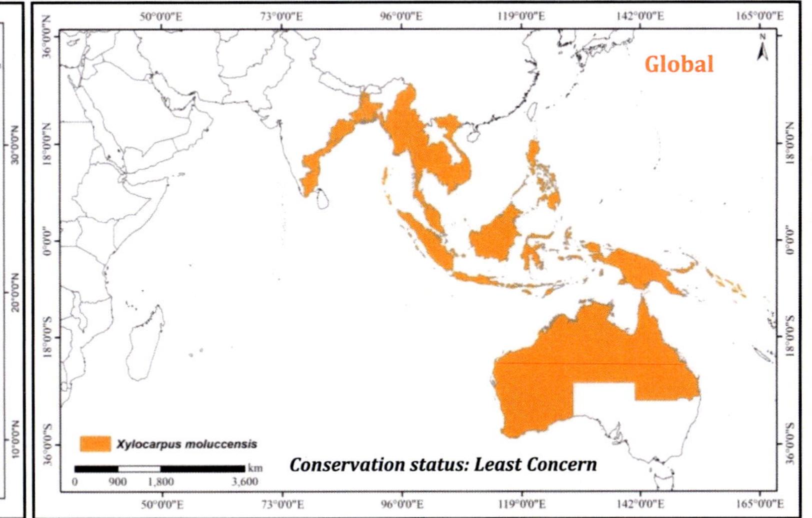

Habitat and Ecology

Common in mid and high intertidal areas and intermediate estuarine position. Occur also in banks of tidal creeks, and along open shores.

Phenology

Flowering April to June; fruiting July to September.

Notes

Identity and nomenclature of this taxon in Indian literature are highly variable. *Xylocarpus mekongensis* and Parkinson's *Carapa* sp., and *Xylocarpus gangeticus* are synonyms of *X. moluccensis.*

Morphological features of *Xylocarpus moluccensis*. (A) Habit; (B) leaf with three pairs of leaflets; (C)leaflet with pointed tip; (D) flower with petal and fused stamens; (E) style and ovary; (F) inflorescences; (G) fruits; (H) seeds; (I) peg like pneumatophores; (J) flaky bark.

Xylocarpus rumphii (Kostel.) Mabb., Malaysian Forester 45(3): 450 (1982).

Meliaceae

Xylocarpus rumphii is a coastal plant and often misidentified mangrove species. It normally grows in high water on cliffs, rocks and sandy uplands. In A & N Islands it has been observed on the rocky shores of Havelock Island, Neil Island, Corbyn's Cove, Chidiyatapu and Little Andaman. It is easily recognized in the field by its compound leaves with 2–4 pairs of heart shaped leaflets, finely fissured bark without any prominent above ground roots and small green fruits.

Inflorescences

KEY CHARACTERS

- Tree grows up to 10m in height, no above ground roots
- Bark is finely fissured grayish, inner bark bright pink to red
- Leaves are ovate to heart shaped, with pointed tip, compound in structure with 2-4 pairs of per leaflet, dark green and veins are prominent on both the sides
- Inflorescences axial with many flowers similar to *X. granatum* and *X. moluccensis*
- The disc at base of ovary is orange in colour
- Fruits are green and smaller than *X. moluccensis* with 4–8 seeds

Species feature

Heart shaped leaflets

Small green color fruits

Flower with orange colour disc at the base of ovary

Brown fissured bark

Xylocarpus rumphii (Kostel.) Mabb

Taxonomy

Tree: spreading, height to 10 m, multi-stemmed, evergreen (A). *Bark*: brown, deeply fissured (F) *Roots*: not above ground, stem base simple. *Leaves*: compound, alternate, 10–15 cm long; petiole 5 cm long; leaflets 2–4 pairs, cordate (H), apex acuminate, base rounded (B), 5–8 × 3–5 cm, ratio of length to width <2, leaflet stalk 0.3–0.5 cm long. *Inflorescence*: axillary, multi-flowered panicle (C), monoecious, male and female flowers present in same inflorescence; peduncle 0.2–0.5 cm long. *Mature flower bud*: 0.3–0.5 cm long, tetramerous, white; calyx lobes 4, yellowish green, united below, 0.2 cm long; petals 4, creamy white, 0.3 × 0.1 cm, oblong (D); stamens united as staminal tube, creamy white (D), 0.3 cm long, upper margin bears 8 erect lobes; style 0.3 cm long, seated on raised ovary, stigma disc shaped; ovary superior, disc shaped, red in colour (E); in male flowers ovary is slender and non-functional, in female flowers stamens are non-functional. *Mature Fruits*: woody, 4 chambered, green (G, I), 7–10 cm in diameter; seeds 4–10, angular, smooth, tetrahedral, pale brown (J), up to 4–8 cm long.

Distribution

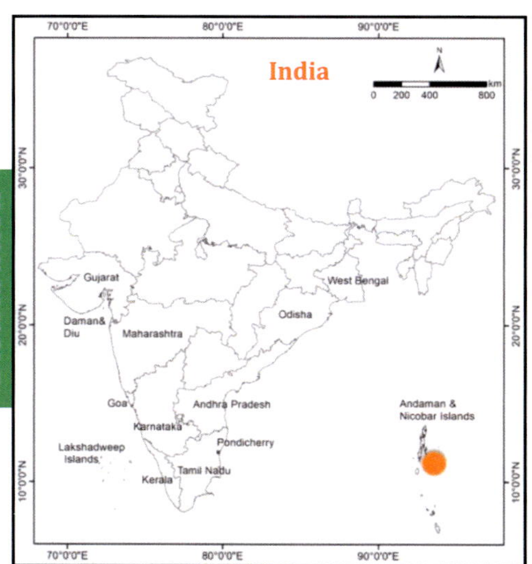

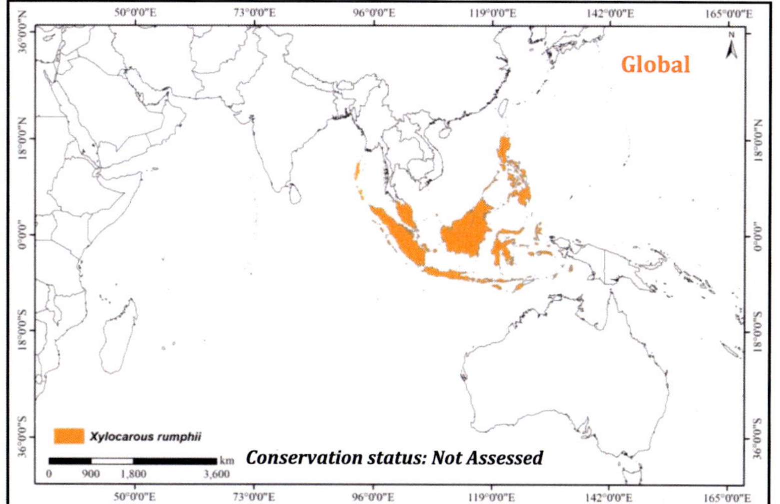

Habitat and Ecology

Often found in exposed shores, rocky cliffs, often near the surf, and sandy substrates above the high-water mark.

Phenology

Flowering June to August; Fruiting September to January.

Notes

This taxon is a non-mangrove species, identified as *X. moluccensis* in most of the earlier literature. Parkinson's *Carapa moluccenis,* found on the rocky coast of the Andaman Islands, is a synonym to *X. rumphii* (Ragavan et al. 2015b)

Morphological features of *Xylocarpus rumphii*. (A) Habit; (b) leaf with three pairs of leaflets; (C) inflorescences; (D) flower with fused stamens.

Mangrove Associate plant species

Mangrove associate plant species are terrestrial plants often extend their distributions into back-mangrove environments (Tomlinson 1986). Since extant mangrove species are descendants of terrestrial angiosperms, mangrove associate species suggesting their predisposition to become adapted to salt-influenced habitats and perhaps representing the initial stages of formation of new mangrove lineages (Ricklefs et al. 2006). Tomlison (1986) listed 48 mangrove associate species in the Indo-West Pacific region. Dagar (2003) listed 188 species of associates, including climbers, epiphytes, and semi-parasites, in Indian mangroves. However, there is no consensus on definition and species composition of mangrove associates . In the recent past Wang et al. (2010) attempted to distinguish true mangroves and mangrove associates more scientifically using leaf characters and the salt content of leaves, but the results did not remove uncertainty. Here we listed few mangrove associate species which were found more frequently in Indian Mangroves during our field surveys.

Ardisia elliptica
Atlantia correa
Barringtonia acutangula
Barringtonia racemosa
Barringtonia asiatica
Caesalpinia crista
Caesalpinia boundac
Calophyllum inophyllum
Carallia brachiata
Cerbera manghas
Cerbera odollum
Clerodendrum inerme
Crinum defixum
Dalbergia spinosa
Derris heterophylla
Derris scandens
Derris trifoliata
Fimbristylis ferruginea
Finlaysonia obovata
Flagellaria indica

Hibiscus tiliaceous
Hoya parasitica
Intsia bijuga
Ipomoea pes-caprae
Ipomoea tuba
Morinda citrifolia
Mucuna gigantea
Myriostachya wightiana
Pandanus fascicularis
P. foetidus
P. furcatus
P. tectoris
Pongamia pinnata
Porteresia coarctata
Pluchea indica
Salacia chinensis
Salicornia brachiata
Salvadora persica
Sarcolobus carinatus
Sarcolobus globosus

Scaevola taccada
Scirpus littoralis
Sesuvium potulacastrum
Suaeda fruticosa
Suaeda maritima
Suaeda monoica
Suaeda nudiflora
Thespesia populnea
Tylophora tenuis

Part 4: State wise glimpses of Indian mangroves

Rhizophora apiculata

West Bengal

In West Bengal, mangroves are present in Sundarbans, the large deltaic complex of three major rivers, the Ganges, Brahmaputra and Meghna, and 62% of total area is shared by Bangladesh and 38% by India. The Sundarban Biosphere Reserve (SBR), at the confluence of the Bay of Bengal covering an area of 9,630 km^2, lies between 21°40′N and 22°40′N latitudes and 88°03′E and 89°07′E longitudes, spreading in districts of South and North 24-Parganas in West Bengal, India. The major portion of SBR is designated as Reserve Forest, covering nearly 4260 km^2 and the area under mangrove vegetation is 2112 km^2 (FSI 2019). Compared to 2017 assessment there has been a decrease of 2 km^2 area of mangroves. The mangroves and their associates in the SBR exhibit great generic and species diversity and it is abode to rare and highly endangered flora and fauna, as well as, the only mangrove-tiger (Royal Bengal Tiger - Panthera *Tigris tigris*) kingdom in the world. Recognizing importance and uniqueness, the United Nations Educational, Scientific and Cultural Organization (UNESCO) declared the Indian part of Sundarbans as a 'World Heritage Site' in 1987, and also designated the 'Sundarban Biosphere Reserve' under the UNESCO Man and the Biosphere (MAB) program in 2001. Recently, Sundarbans has been declared as a Ramasar Site, considering ecological significance. Totally 33 true mangrove species belonging to 21 genera and 13 families have been identified in Indian Sundarbans. Among the mangrove species listed, *Rhizophora apiculata*, *Bruguiera parviflora*, *Ceriops decandra*, *Kandelia candel*, *Aglaia cucullata*, *Excoecaria indica*, *Xylocarpus moluccensis*, *Xylocarpus granatum*, *Heritiera fomes*, *Scyphiphora hydrophyllacea*, *Brownlowia tersa*, *Nypa fruticans*, *Acanthus volubilis*, *Dolichandrone spathacea* and *Phoenix paludosa* are under serious threat in Sundarbans (Kathiresan 2008; Gopal and Chauhan 2006). Particularly *Heritiera fomes* (locally called Sundari, from which Sundarbans derives its name), *Nypa fruticans* and *Phoenix paludosa* are declining rapidly due to the reduction of freshwater input (Gopal and Chauhan 2006).

1. *Acanthus ilicifolius*
2. *Acanthus volubilis*
3. *Acrostichum aureum*
4. *Aegialitis rotundifolia*
5. *Aegiceras corniculatum*
6. *Aglaia cucullata*
7. *Avicennia alba*
8. *Avicennia marina*
9. *Avicennia officinalis*
10. *Brownlowia tersa*
11. *Bruguiera cylindrica*
12. *Bruguiera gymnorhiza*
13. *Bruguiera parviflora*
14. *Bruguiera sexangula*
15. *Ceriops decandra*
16. *Ceriops tagal*
17. *Cynometra iripa*
18. *Dolichandrone spathacea*
19. *Excoecaria agallocha*
20. *Excoecaria indica*
21. *Heritiera fomes*
22. *Kandelia candel*
23. *Lumnitzera racemosa*
24. *Nypa fruticans*
25. *Phoenix paludosa*
26. *Rhizophora apiculata*
27. *Rhizophora mucronata*
28. *Scyphiphora hydrophylacea*
29. *Sonneratia alba*
30. *Sonneratia apetala*
31. *Sonneratia caseolaris*
32. *Xylocarpus granatum*
33. *Xylocarpus moluccensis*

Ceriops tagal

Odisha

The total length of coastline in Odisha is about 430 km. The coastal zone is situated between the latitude 19^0 N and 22^0 N and longitude between 85^0E and 87^0 30'E. The mangroves of the Odisha coast are distributed in the Mahanadi, Brahmani and Baitarani Delta i.e., the Bhitarkanika mangroves and Balasore coast. Among these three zones, Bhitarkanika is the most important due to its largest stretch and unique biodiversity. It is also considered as the fourth largest mangrove zone of the country following Sundarbans, Gujarat and Andaman and Nicobar Islands. During 2002 the Bhitarkanika mangroves covering an area of 672 km^2 was declared as a Ramsar site being a wetland of international importance. In Odisha the total area under mangrove vegetation is 251 km^2 (FSI 2019.) Compared to 2017 assessment there has been an increase of 8 km^2 area of mangroves (FSI 2019). Totally 34 true mangrove species belonging to 20 genera and 13 families have been recognized from mangroves of Odisha. Among them, species of rare occurrence are *Acrostichum speciosum*, *Bruguiera parviflora*, *Ceriops tagal*, *Dolichandrone spathacea*, *Sonneratia alba*, *Sonneratia caseolaris*, *Xylocarpus moluccensis* and *Rhizophora apiculata* (Kathiresan 2008). *Heritiera fomes*, *Excoecaria agallocha* and *Avicennia officinalis* are dominant at Bhitarkanika (Mishra et al. 2003). Upadhyay and Mishra (2014) mentioned that *Excoecaria agallocha*, *Avicennia officinalis* and *Hertiera fomes* accounted for more than 50% of the total Importance Value Index (IVI) in Bhitarkanika sanctuary. In Bhitarkanika, there is a great assemblage of genetic diversity of mangroves in an area of about 850 hectares on the isolated island of "Kalibhanj Dia". Mangrove floristics of the Bhitarkanika Wildlife Sanctuary, the Mahanadi delta and Chilika Lake on Odisha coast are well-known, but not for estuaries such as Devi, Budhabalanga, Rushikulya, Subarnarekha, etc. (Panda et al. 2013).

1. *Acanthus ilicifolius*
2. *Acanthus volubilis*
3. *Acrostichum aureum*
4. *Acrostichum speciosum*
5. *Aegialitis rotundifolia*
6. *Aegiceras corniculatum*
7. *Aglaia cucullata*
8. *Avicennia alba*
9. *Avicennia marina*
10. *Avicennia officinalis*
11. *Brownlowia tersa*
12. *Bruguiera cylindrica*
13. *Bruguiera gymnorhiza*
14. *Bruguiera parviflora*
15. *Bruguiera sexangula*
16. *Ceriops decandra*
17. *Ceriops tagal*
18. *Cynometra iripa*
19. *Dolichandrone spathacea*
20. *Excoecaria agallocha*
21. *Excoecaria indica*
22. *Heritiera fomes*
23. *Heritiera littoralis*
24. *Kandelia candel*
25. *Lumnitzera racemosa*
26. *Phoenix paludosa*
27. *Rhizophora apiculata*
28. *Rhizophora mucronata*
29. *Scyphiphora hydrophylacea*
30. *Sonneratia alba*
31. *Sonneratia apetala*
32. *Sonneratia caseolaris*
33. *Xylocarpus granatum*
34. *Xylocarpus moluccensis*

Andhra Pradesh

Total length of Andhra Pradesh coastline is about 1014 km. The mangrove forests in Andhra Pradesh are located in the estuaries of the Godavari and the Krishna rivers, in East Godavari, Krishna and Guntur districts. These are situated within the latitude 13° 30'N and 19° 00'N and longitude 80° 05'E and 85° 05'E. Apart from these, mangroves are also found in small patches along the coast of Visakhapatnam, West Godavari, Guntur and Prakasam districts. Swain et al. (2008) identified two mangrove habitats in Srikakulam districts namely Nuvvalarevu and Bhavanapadu. *Avicennia marina* and *Acanthus ilicifolius* are recorded in Nuvvalarevu, while in Bhavanapadu *Avicennia marina*, *Bruguiera cylindrica*, *Ceriops decandra*, *Sonneratia apetala* and *Acanthus ilicifolius* are recorded. Area under mangrove vegetation is 404 km^2 (FSI 2019). Totally 22 true mangrove species belonging to 15 genera and 10 families have been recognized in Andhra Pradesh. Among them, the species of rare occurrence are *Aegialitis rotundifolia*, *Scyphiphora hydrophyllacea*, *Sonneratia alba* and *Xylocarpus granatum*. *Brownlowia tersa* is reported from Ramannapalem of East Godavari, which forms an addition to the mangrove flora of Andhra Pradesh (Venu et al. 2006; Swain and Rama Rao 2008). Members of the genus *Avicennia* are dominant in mangrove habitats of Andhra Pradesh (Arisdason et al. 2008; Swain et al., 2008; Nabi et al. 2012; Nabi and Rao 2012). Coringa Wildlife Sanctuary, a part of the Godavari mangroves was declared as a sanctuary in July 1978 to conserve the mangrove vegetation of the estuary, extending in an area of about 235 sq.km. It is located between 16^{0}30' to 17^0-00' N latitude and 82^0-14' to 82^0-23' E longitude in the East Godavari District of Andhra Pradesh, India. The Sanctuary has a unique distinction of having an 18 Km long sand spit in the North Eastern side, where the species of "Olive Ridley sea turtle" (endangered species) nests during January-March of every year. *Scyphiphora hydrophyllacea* is restricted to nearer to Sacramento light house in Kandikuppa.

1. *Acanthus ilicifolius*
2. *Acrostichum aureum*
3. *Aegialitis rotundifolia*
4. *Aegiceras corniculatum*
5. *Avicennia alba*
6. *Avicennia marina*
7. *Avicennia officinalis*
8. *Brownlowia tersa*
9. *Bruguiera cylindrica*
10. *Bruguiera gymnorhiza*
11. *Ceriops decandra*
12. *Ceriops tagal*
13. *Excoecaria agallocha*
14. *Kandelia candel*
15. *Lumnitzera racemosa*
16. *Rhizophora apiculata*
17. *Rhizophora mucronata*
18. *Scyphiphora hydrophylacea*
19. *Sonneratia alba*
20. *Sonneratia apetala*
21. *Xylocarpus granatum*
22. *Xylocarpus moluccensis*

Red colour flower of *Lumnitzera littorea*

Tamil Nadu

The coastline of Tamil Nadu extends to about 950 km, within the latitude 8^0 N-13^0 30'N and longitude 77^0 15'E – 80^0 20'E with about 46 big and small rivers. All these rivers carry freshwater and silt particles from the upper reaches and discharge in the coastal zone (Sudhakar Reddy 2008). In Tamil Nadu, mangroves are confined to Pichavaram, Muthupet and Gulf of Mannar. Area under mangrove vegetation is 45 km^2 (FSI 2019). Compared to 2017 assessment there has been a decrease of 4 km^2 area of mangroves (FSI 2019). Totally 17 true mangrove species belonging to 12 genera and 8 families have been recognized from Tamil Nadu. Among them, the species of restricted occurrence are *Rhizophpora* x *annamalayana*, *Ceriops tagal*, *Sonneratia apetala*, *Xylocarpus granatum*, *Xylocarpus moluccensis*, *Acrostichum aureum* and *Lumnitzera racemosa* (Kathiresan 2008). Compared to the earlier reports, three species i.e., *Sonneratia apetala*, *Xylocarpus granatum* and *Acrosticum aurem* could not be traced by Arunprasath and Gomathinayagam (2014). Pichavaram is well known mangrove area of Tamil Nadu. Situated in latitude of 11° 27'N and longitude of 79° 47' E about 250 km south of the city of Madras, on the south east coast of India. It is located in the Vellar-Coleroon estuarine complex and has many islands separated by intricate waterways. Totally the mangrove covers an area of 12.05 km^2. It is traversed by a large number of channels and creeks which connect the Coleroon estuary in the south and Vellar estuary in the north. Muthupet is the largest mangrove forest in Tamil Nadu, as a part of Point Calimere Wildlife Sanctuary. The Muthupet has the second largest lagoon system in Tamil Nadu. It is located in the southernmost end of the Cauvery delta along the Palk Strait, a part of the Great Vedaranyam swamp. This is the only Ramsar Site in Tamil Nadu. Monospecific dominance of the mangrove species *Avicennia marina* is a unique feature of that mangrove area.

1. *Acanthus ilicifolius*
2. *Acrostichum aureum**
3. *Aegiceras corniculatum*
4. *Avicennia marina*
5. *Avicennia officinalis*
6. *Bruguiera cylindrica*
7. *Ceriops decandra*
8. *Ceriops tagal*
9. *Excoecaria agallocha*
10. *Lumnitzera racemosa*
11. *Pemphis acidula*
12. *Rhizophora × annamalayana*
13. *Rhizophora apiculata*
14. *Rhizophora mucronata*
15. *Sonneratia apetala* *
16. *Xylocarpus granatum**
17. *Xylocarpus moluccensis*

* Doubtful

Inflorescence of *Rhizophora × lamarckii*

Kerala

Rare occurrence of petals in *Sonneratia ovata*

Mangrove patches of Kerala are mainly distributed in intertidal areas of Kochi, Vembanad, Kollam, Thiruvananthapuram, Kannur, Kozhikode and Kottayam. Area under mangrove vegetation in Kerala is 9 km^2 (FSI 2019). Totally 19 species belonging to 12 genera and 8 families have been recognized as true mangroves in Kerala. The earliest reference on mangrove floristics in Kerala (Van Rheede 1678–1693) reported eight mangrove species: *A. corniculatum*, *A. officinalis*, *E. agallocha*, *K. candel*, *Lumnitzera racemosa*, *Rhizophora cylindrica* (= *R. apiculata*), and *R. mucronata*. Subsequently Drury (1864) had described a few more plants, apart from the ones listed by Van Rheede (1678–1693) namely, *Eriops candolleanus* (= *C. tagal*) and *Bruguiera eriopetala* (= *Bruguiera sexangula*). In the past there were many studies on the occurrence of mangrove flora along the Kerala coasts (Troup 1921; Govinda-Menon 1930; Erlanson 1936; Mudailarm and Kamath 1954; Thomas 1962; Rao and Sastry 1974; Blasco 1975; Kurian 1980; Ramachandran et al. 1986; Ramachandran and Mohanan 1987; Ramachandran and Mohanan 1990; Basha 1991, 1992; Anupama and Sivadasan 2004). *Avicennia* species are dominant in mangrove habitats of Kerala (Vidyasagaran and Madhusoodanan 2014; Sreelakshmi et al. 2018). Kathiresan (2008) reported the occurrence of *Acanthus ebracteatus* from Kerala but it was not recorded in studies of Vidyasagaran et al. (2011), Ram and Shaji (2013), and Sindhumathi et al. (2014) in recent times. Reports of *Avicennia alba* from Kerala is erroneous (e.g. Sreelakshmi et al. 2018). The mangrove species of restricted occurrence are *Excoecaria indica*, *Bruguiera sexangula*, *Sonneratia alba*, *Sonneratia caseolaris*, *Aegiceras corniculatum*, *Lumnitzera racemosa*, and *Rhizophora apiculata* (Kathiresan 2008).

1. *Acanthus ilicifolius*
2. *Acrostichum aureum*
3. *Aegiceras corniculatum*
4. *Avicennia marina*
5. *Avicennia officinalis*
6. *Bruguiera cylindrica*
7. *Bruguiera gymnorhiza*
8. *Bruguiera sexangula*
9. *Ceriops tagal*
10. *Dolichandrone spathacea*
11. *Excoecaria agallocha*
12. *Excoecaria indica*
13. *Kandelia candel*
14. *Lumnitzera racemosa*
15. *Nypa fruticans*
16. *Rhizophora apiculata*
17. *Rhizophora mucronata*
18. *Sonneratia alba*
19. *Sonneratia caseolaris*

Karnataka

Karnataka state lies between latitude 11^0 12' N and 18^0 12'N and longitude 73^0 48' E - 78^0 18' E. The coastline is about 320 km long and is overshadowed by the Western Ghats mountain range. The coast is lined with sandy and occasionally rocky shores. Mangroves in Karnataka are of fringing type and are confined to Uttar Kannada and Dakshina Kannada. Area under mangroves in Karnataka is 10 km^2 (FSI 2019). The important estuarine areas where mangroves are present in Dakshina Kannada are Netravathi-Gurupur, Mulki-Pavanje, Udayavara-Pangala, Swarna-Sita-Kodi, Chakra-Haladi-Kollur, Baindur hole and Shiroor hole while in Uttarakannda the mangroves are present in the Honovar, Venkatapur, Sharavathi, Aghanashini, Gangavali and Kali river estuarine complexes. Totally 16 species belonging to 11 genera and 7 families have been recognized as true mangrove species in Karnataka. Earlier mangrove floristics of Karnataka was studied by Arora and Agarwal (1960), Rao and Suresh (1990), Singh et al. (1999), Nayak and Andrade (2008), Sulochanan (2010). *Bruguiera cylindrica*, *Lumnitzera racemosa* and *Acrostichum aureum* are recently recorded by Nayak and Andrade (2008) from Karnataka. Species of rare occurrences are *Bruguiera parviflora*, *Sonneratia caseolaris*, *Aegiceras corniculatum* and *Rhizophora apiculata* (Kathiresan 2008). The coast of Karnataka is predominantly occupied by *Rhizophora mucronata* followed by *Avicennia officinalis*, *Sonneratia apetala* and *Kandelia candel*. The mean density of trees is 1,740.6 per hectare (Bhatt and Kathiresan 2010).

1. *Acanthus ilicifolius*
2. *Acrostichum aureum*
3. *Aegiceras corniculatum*
4. *Avicennia alba*
5. *Avicennia marina*
6. *Avicennia officinalis*
7. *Bruguiera cylindrica*
8. *Bruguiera gymnorhiza*
9. *Ceriops decandra*
10. *Excoecaria agallocha*
11. *Kandelia candel*
12. *Lumnitzera racemosa*
13. *Rhizophora apiculata*
14. *Rhizophora mucronata*
15. *Sonneratia alba*
16. *Sonneratia caseolaris*

Woody calyx of *Sonneratia griffithii*

Goa

The total length of the Goa coast is approximately 120 km and within the latitude 15^0 00'N - 15^0 52'N and longitude 73^0 30'E - 74^0 44'E. In Goa, the mangroves are present in Mandovi estuary, Zuari estuary and Cumbarjua Canal. In addition, other parts of Galgibag, Talpona, Sal, Chapora and Terekhol river mouths also are endowed with mangrove vegetation. Chodan (Chorao) Island is the totally protected mangrove area of 160 ha., near the village of Chodan (Chorao). This mangrove forest has been declared as Dr. Salim Ali Bird Sanctuary. It harbours good mangrove flora. The whole sanctuary area consists of mangrove vegetation interspersed with water channel, having tidal variations. The area is very rich in marine fauna and avifauna. Area under mangrove vegetation is 26 km^2 (FSI 2019). Goa has 16 true mangrove species belonging to 11 genera and 7 families. Earlier floristics of mangrove of Goa was studied by Untawale et al. (1973), (1982); Untawale (1986), Dwivedi et al. (1974), Rao (1985), (1986); Naskar and Mandal (1999), Jagtap (1985), Kothari and Rao (2002). In recent past Anant (2012) reported only six true mangrove species from Zuari estuary. Dagar and Singh (1999) reported the occurrence of *Sonneratia apetala* and *Xylocarpus granatum*, but these could not be recollected in the survey made by Kothari and Rao (2002). Jagtap (1985) reported *Bruguiera parviflora* from Mandovi estuary and later it was confirmed that *Bruguiera cylindrica* was reported as *Bruguiera pariviflora* by him (Naskar and Mandel 1999). *Kandelia candel*, a rare species on the west coast, is found in plenty in Mandovi, Mapusa and Zuari rivers. However, *Avicennia* species is dominant in most of the places with high salinity and *Sonneratia* species is dominant in low salinity areas. Species of rare occurrence are *Ceriops tagal*, *Bruguiera cylindrica* and *Lumnitzera racemosa* (Kathiresan 2008). Mandovi River is one of the best mangrove forests and housing most of the species found in Goa.

1. *Acanthus ilicifolius*
2. *Acrostichum aureum*
3. *Aegiceras corniculatum*
4. *Avicennia alba*
5. *Avicennia marina*
6. *Avicennia officinalis*
7. *Bruguiera cylindrica*
8. *Bruguiera gymnorhiza*
9. *Ceriops tagal*
10. *Excoecaria agallocha*
11. *Kandelia candel*
12. *Lumnitzera racemosa*
13. *Rhizophora apiculata*
14. *Rhizophora mucronata*
15. *Sonneratia alba*
16. *Sonneratia caseolaris*

Purple colour flowers of *Acanthus ilicifolius*

Maharashtra

The total length of the Maharashtra coast is 720 km long, characterised by several pocket beaches flanked by rocky cliffs. The Maharashtra coastal zone extends between the latitude 15 52'N and 20 10'N and longitude 72 10'E and 73 10'E and falls under five districts (Sindhudurg District, Ratnagiri District, Thane District, Bombay (Mumbai) District, Raigad District). Mangroves are mainly concentrated at the mouths of rivers, the Vashisthi, the Savitri, the Kundalika, the Dharmatar, the Panvel, the Vasai, the Thane and the Vaitarana. In Mumbai region, mangroves are spreading along the tidal river creeks and backwaters of Achara, Deogadh, Vijaydurg, Ratnagiri, Kundalica and Mumbradiva. Others include Veldur, Vikroli, Shreevardhan, Vaitarna, Vasai-Manori and Malvan. Area under mangrove vegetation is 320 km^2 (FSI 2019). Compared to 2017 assessment there has been an increase of 16 km^2 area of mangroves (FSI 2019). Totally 22 mangrove species belonging to 15 genera and 10 families have been recorded. Earlier mangrove floristics of Maharashtra coast was studied by Cooke (1901–1908), Blatter (1905), Bharucha and Navalkar (1950), Navalkar (1951), Satyanarayana (1958), Shah (1962). Bhosale et al. (2002) reported the occurrence of *Xylocarpus granatum, Dolichandrone spathacea* and *Cynometra iripa* and Shaikh et al. (2011) reported *Heritiera littoralis* from Maharashtra. In recent past Yeragi and Yeragi (2014) listed 20 mangrove species including three non mangrove species i.e *Salvadora persica*, *Derris trifoliata*, *Derris scandens*. Earlier Jagtap et al. (1994) reported 20 mangrove species and in his list *Bruguiera cylindrica* was identified as *Bruguiera parviflora. Avicennia marina* is dominant in mangroves of Maharashtra and the species of rare occurrence are *Cynometra iripa*, *Heritiera littoralis*, *Dolichandrone spathacea*, *Sonneratia caseolaris* and *Xylocarpus granatum* (Kathiresan 2008)

Ribbon like petals of *Sonneratia alba*

1. *Acanthus ilicifolius*
2. *Acrostichum aureum*
3. *Aegiceras corniculatum*
4. *Avicennia alba**
5. *Avicennia marina*
6. *Avicennia officinalis*
7. *Bruguiera cylindrica*
8. *Bruguiera gymnorhiza*
9. *Bruguiera parviflora**
10. *Ceriops tagal*
11. *Cynometra iripa*
12. *Dolichandrone spathacea*
13. *Excoecaria agallocha*
14. *Heritiera littoralis*
15. *Kandelia candel*
16. *Lumnitzera racemosa*
17. *Rhizophora apiculata*
18. *Rhizophora mucronata*
19. *Sonneratia alba*
20. *Sonneratia apetala*
21. *Sonneratia caseolaris*
22. *Xylocarpus granatum*

* Doubtful

Flower of *Sonneratia griffithii*

Gujarat

In Gujarat mangroves are present in Indus deltaic region (Kachchh) i.e., Kori creek and Sir Creek area, Gulf of Kachchh, South Gujarat and Gulf of Khambhat. Totally 15 species belonging to 10 genera and 7 families have been recognized as true mangrove species of Gujarat. Area under mangrove vegetation is 1177 km^2, and there has been an increase of 37 km^2 in comparison with 2017 assessment (FSI 2019). Among the 15 species one species i.e., *Rhizophora apiculata* reported earlier, (Singh 1994; Kothari and Singh 1998; GEER 2000) has not been recorded for more than one decade, leaving the mangrove diversity of Gujarat to be 14 (Pandey and Pandey 2010). Area-wise mangroves of Gujarat stand second in position in India but there is less diversity compared to other habitats of India. *Avicennia marina* is dominant in mangrove habitats of Gujarat. Kachchh which holds the largest mangrove cover in the state is largely represented by only one species, *Avicennia marina*. Gulf of Kachchh has four species i.e., *Avicennia marina*, *Aegiceras corniculatum*, *Rhizophora mucronata* and *Ceriops tagal* and it represents 15.2% of Gujarat mangrove cover. Gulf of Khambhat has *Avicennia marina*, *Avicennia officinalis*, *Acanthus ilicifolius* and *Sonneratia apetala* and represents 10.1% of Gujarat's mangrove cover. However, South Gujarat with only 0.6% of mangrove cover hosts 14 mangrove species. Recently Bhatt et al. (2009) reported the occurrence seven mangrove species in Purna estuary, South Gujarat, which makes it one of the most diverse mangrove patches in the State. The species of rare occurrence in Gujarat are *Ceriops tagal*, *Ceriops decandra*, *Kandelia candel*, *Excoecaria agallocha*, *Lumnitzera racemosa*, *Rhizophora mucronata*, *Aegiceras corniculatum*, *Sonneratia apetala*, *Bruguiera gymnorhiza* and *Bruguiera cylindrica* (Kathiresan 2008).

1. *Acanthus ilicifolius*
2. *Aegiceras corniculatum*
3. *Avicennia alba*
4. *Avicennia marina*
5. *Avicennia officinalis*
6. *Bruguiera cylindrica*
7. *Bruguiera gymnorhiza*
8. *Ceriops decandra*
9. *Ceriops tagal*
10. *Excoecaria agallocha*
11. *Kandelia candel*
12. *Lumnitzera racemosa*
13. *Rhizophora apiculata* *
14. *Rhizophora mucronata*
15. *Sonneratia apetala*

* Doubtful

Flower of *Rhizophora* sp

Stem base of *Sonneratia alba*

Andaman and Nicobar Islands

Andaman and Nicobar Islands (A & N Islands) is a group of 572 Islands and Islets, located in the Bay of Bengal, off the eastern coast of India. Its irregular and deeply indented coastline results in innumerable creeks, bays and estuaries and these facilitate the development of extensive and luxuriant growth of mangrove forests with a high degree of biodiversity. The mangroves of A & N Islands are probably the best in India in terms of density and growth (Ragavan et al. 2019). According to the latest estimate by the Forest Survey of India (FSI 2019), 616 km^2 is under mangroves in A & N Islands; out of which, 614 km^2 is in Andaman Islands with just 2 km^2 spread in Nicobar Islands. Compared to 2017 assessment there has been a decrease of 1 km^2 area of mangroves (FSI 2019). The resulting upliftment and submergence in the North Andaman and Southern group of Islands, respectively, is leading to gradual death of mangroves and replacement by non-mangrove vegetation. Our extensive survey revealed the occurrence of 38 species belonging to 12 families and 19 genera. Recently reported mangrove species are *Sonneratia ovata*, *Sonneratia lanceolata*, *Sonneratia × gulngai*, *Sonneratia × urama* and *Excoecaria indica* (Goutham Bharati et al. 2012; Ragavan 2015). Species of rare occurrence are *Acanthus ebracteatus*, *Acanthus volubilis*, *Acrosticum speciosum*, *Aegiceras corniculatum*, *Brownlowia tersa*, *Excoecaria indica*, *Lumnitzera racemosa*, *Pemphis acidula*, *Rhizophora × annamalayana*, *Rhizophora × lamarckii*, *Rhizophora stylosa*, *Sonneratia caseolaris*, *Sonneratia griffithii*, *Sonneratia lanceolata*, *Sonneratia ovata*, *Sonneratia × urama*, *Sonneratia × gulngai* and *Xylocarpus moluccensis*.

1. *Acanthus ebracteatus*
2. *Acanthus ilicifolius*
3. *Acanthus volubilis*
4. *Acrostichum aureum*
5. *Acrostichum speciosum*
6. *Aegiceras corniculatum*
7. *Avicennia marina*
8. *Avicennia officinalis*
9. *Brownlowia tersa*
10. *Bruguiera cylindrica*
11. *Bruguiera gymnorhiza*
12. *Bruguiera parviflora*
13. *Ceriops tagal*
14. *Cynometra iripa*
15. *Dolichandrone spathacea*
16. *Excoecaria agallocha*
17. *Excoecaria indica*
18. *Heritiera littoralis*
19. *Lumnitzera littorea*
20. *Lumnitzera racemosa*
21. *Nypa fruticans*
22. *Pemphis acidula*
23. *Phoenix paludosa*
24. *Rhizophora × annamalayana*
25. *Rhizophora × lamarckii*
26. *Rhizophora apiculata*
27. *Rhizophora mucronata*
28. *Rhizophora stylosa*
29. *Scyphiphora hydrophylacea*
30. *Sonneratia × gulngai*
31. *Sonneratia × urama*
32. *Sonneratia alba*
33. *Sonneratia caseolaris*
34. *Sonneratia griffithii*
35. *Sonneratia lanceolata*
36. *Sonneratia ovata*
37. *Xylocarpus granatum*
38. *Xylocarpus moluccensis*

Greenish white calyx of *Bruguiera gymnorhiza*

Puducherry, Lakshadweep and Diu & Daman

A total of 15 true mangrove species 10 genera and 7 families have been recognized in Union territory of Puducherry. Of the four regions, mangrove species diversity is rich in Yanam with 14 true mangrove species (Balachandran et al. 2009). Saravanan et al. (2008) reported 7 true mangrove species from Puducherry. In Puducherry, mangroves are present in three villages namely, Ariankuppam, Murungapakkam, Veerampattinam and two islets - Thengaithittu and Ashramthittu. Area under mangrove vegetation in Puducherry is 2 km^2 (FSI 2019).

The Lakshadweep comprises 36 islands. Mandal and Naskar (2008) reported only *Bruguiera parviflora* from Minicoy Island. Earlier Nasser et al. (1999) reported *Bruguiera cylindrica*, *Ceriops tagal* and single solitary tree of *Avicennia marina* from Minicoy Island. However, based on recent floristics survey (unpublished data), mangroves of Minicoy Island consist of three mangrove species viz. *Bruguiera cylindrica, Ceriops tagal* and *Pemphis acidula*. Of these homogenous patch of *B. cylindrica* is in the south eastern side, whereas mixed strand of *Ceriops tagal* and *Pemphis acidula* is present in the south western side. In addition, *Pemphis acidula* is also observed along entire coastal length of Minicoy Island. Since Lakshadweep Islands are coral atolls, the sediments often consist of coral deposits which favour the growth of *Pemphis acidula*. The single solitary tree of *A. marina* reported by Nasser et al. (1999) has not been found in the recent survey.

In Diu & Daman, mangroves are present at the banks of Chasi river estuary. Totally four mangrove species namely *Avicennia marina*, *Acanthus ilicifolius*, *Aegiceras corniculatum* and *Sonneratia apetala* have been recorded from Diu & Daman (Sharma and Sikarvar 2014). Area under mangrove vegetation in Diu & Daman is 3 km^2 (FSI 2019).

Puducherry

1. *Acanthus ebracteatus*
2. *Acanthus ilicifolius*
3. *Aegiceras corniculatum*
4. *Avicennia alba*
5. *Avicennia marina*
6. *Avicennia officinalis*
7. *Bruguiera cylindrica*
8. *Bruguiera gymnorhiza*
9. *Ceriops tagal*
10. *Excoecaria agallocha*
11. *Lumnitzera racemosa*
12. *Rhizophora apiculata*
13. *Rhizophora mucronata*
14. *Sonneratia apetala*
15. *Xylocarpus moluccensis*

Lakshadweep

1. *Bruguiera cylindrica*
2. *Ceriops tagal*
3. *Pemphis acidula*

Diu & Daman

1. *Acanthus ilicifolius*
2. *Aegiceras corniculatum*
3. *Avicennia marina*
4. *Sonneratia apetala*

Part 5: Significance of Mangroves

Mangrove forests are ecologically significant and economically important (Kathiresan 2018). They provide ecosystem services worth at least US$ 1.6 billion each year and support coastal livelihoods worldwide (Costanza et al. 1997). They serve as the nursery, feeding and breeding grounds for crabs, prawns, molluscs, finfish, birds, reptiles and mammals. A large amount of global fish catches (up to 80%) is dependent on mangroves, thereby ensuring the food security of coastal people (Ellison 2008). The mangroves provide firewood, timber, cattle feed, honey, medicines and tourism development. They protect groundwater aquifers from seepage of seawater, thereby ensuring water security for coastal population. Mangrove forests remove coastal pollution particularly toxic heavy metals. They offer coastal protection against the fiery effects of natural calamities such as tsunami, storm surges, cyclone and floods (Kathiresan and Bingham 2001; Kathiresan and Qasim 2005; Kathiresan and Rajendran 2005). The potential of mangroves in carbon capture and sequestration is remarkable in mitigating the impacts of global warming and climate change (Kathiresan et al. 2013, 2014).

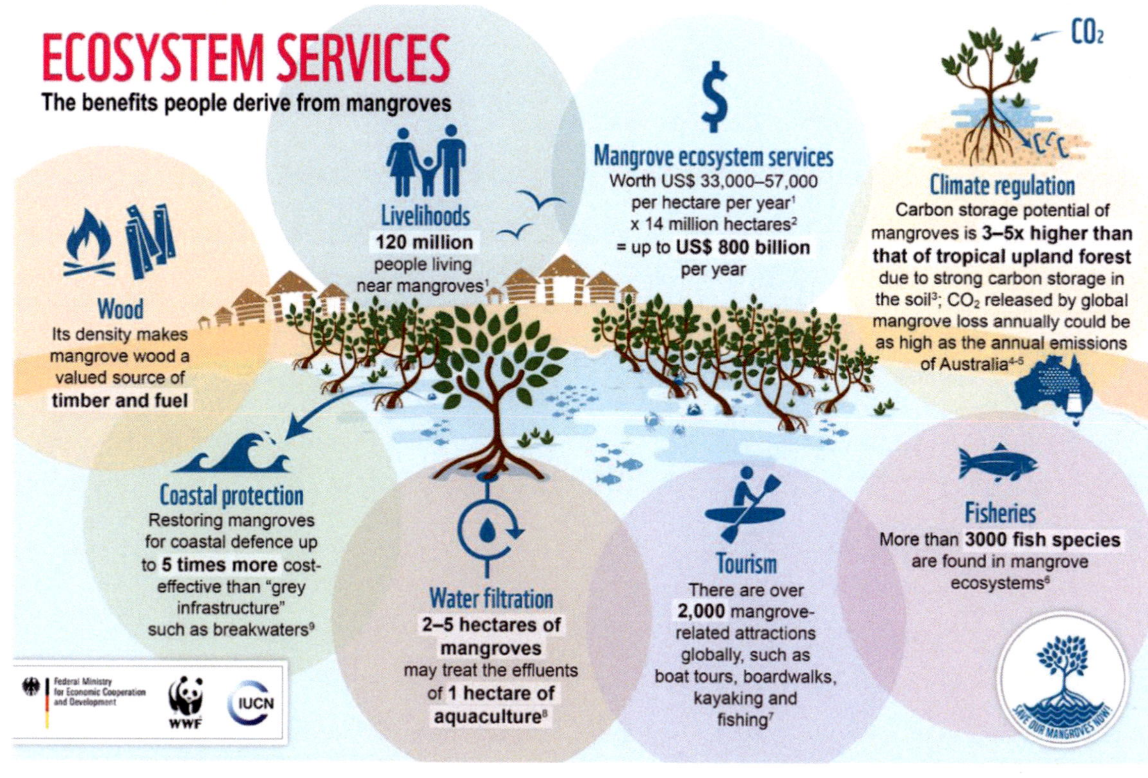

Sources: ❶ UNEP, 2014 • ❷ Giri et al., 2011 • ❸ In the Indo-Pacific region: Donato et al., 2011 • ❹ Up to 450 million t CO_2: Pendleton et al., 2012 • ❺ In 2015: EDGARv4.3.2., 2018 • ❻ Sheaves, 2017 • ❼ Spalding et al., 2016 ❽ Primavera et al., 2007 • ❾ In Vietnam: Narayan et al., 2016

Biodiversity

Mangroves support rich biodiversity. The faunal communities in mangroves chiefly consist of terrestrial animals (insects, birds, mammals, and reptiles), aquatic animals (fish, crustaceans, mollusks, and echinoderms), and benthic animals (polychaetes, brachyuran crabs, wood-boring animals, mud-burrowing bivalves, gobiid fish, and gastropods) and sessile bivalves, such as oysters, Modiolus spp., and barnacles (Kathiresan et al. 2015). People are living inside the mangrove forests of Sundarbans and A & N Islands. The tribal people mainly the Onge, Jarawas, Sentinelese, Great Andamanese, Shompen, and Nicobarese are mainly dependent on mangrove forest resources (Dagar and Dagar 1986 1991, 1999; Dagar and Singh 1999). In addition to these macroforms, the mangroves harbour a diverse group of microorganisms including bacteria, fungi, cyanobacteria, microalgae, and fungi-like protists. In tropical mangroves, bacteria and fungi constitute 91% of the total microbial biomass, whereas algae and protozoa represent only 7% and 2% of it respectively (Alongi 1988). India has the highest record of biodiversity in mangrove forests of the world, and no other countries have recorded so many species to be present in the ecosystem. So far, 5747 species including 16% of flora and 84% of faunal species have been recorded in Indian mangrove forests (Kathiresan 2018).

Faunal diversity of Indian mangroves (Source Kathiresan 2018)

	Groups	Number of species
Faunal diversity	Prawns and lobsters	55
	Crabs	145
	Insects	661
	Mollusks	337
	Other Invertebrates	745
	Fish Parasites	7
	Fins Fish	554
	Amphibians	13
	Reptiles	84
	Birds	513
	Mammals	68
	Total Faunal diversity	3182
Microbial diversity	Bacteria	69
	Fungi	103
	Actinomycetes	23
	Lichens	32
	Total Microbial diversity	227

Nursery habitat

Mangrove forests are valuable nursery areas for crustaceans, mollusks, and fishes as they provide a rich source of food for growth and shelter for protection from predators. Nearly 80% of the fish catches are directly or indirectly dependent on mangrove and other coastal ecosystems worldwide (Kathiresan 2015). Aerial roots of the mangroves provide a protected habitat for larvae and early juveniles. Mangrove litter fall forms the source for the detrital food web on which many fish depend. The most common fishes in Indian mangrove waters are species of *Liza, Mugil, Lates, Lutjanus, Hilsa, Etroplus suratensis* etc. A special group of fish species found abundantly in the mangroves is the mudskippers and these are physiologically and morphologically adapted to an amphibious existence in intertidal area with variable environmental conditions. Prawns that are commonly occurring in the mangroves are species of *Penaeus* and *Metapenaeus* while the crabs are mainly *Scylla serrata*. The molluscans of mangroves are mainly represented by *Crassostrea madrasensis, Gelonia* sp. and the gastropod *Cerithedia* sp. In addition, juveniles of coral reef fish (e.g. Serranidae, Lutjanidae, Lethrinidae, etc.) are dependent on mangroves for shelter and food. A large number of marine fish and shrimps complete their early stages of life cycle in mangroves (Kathiresan 2015).

Telescopium telescopium

Uca spp.,

Mollusks

Coastal Protection

Protective role of mangroves in disaster risk reduction is widely recognized. Mangroves prevent coastal erosion by reducing the height and energy of wind and swell waves passing through them and minimize the impact of natural hazards such as typhoons, cyclones, hurricanes, and tsunamis. A large tract of mangroves is needed to significantly reduce waves, storm surges and flooding impacts. For example, 100 m of mangroves is required for 13-66% reduction of wave height, and similarly for reduction of storm surge height of 5-50 cm, one kilometre of mangrove forest is needed. Mangroves saved human lives and properties during the Indian Ocean tsunami of 2004 by reducing the destructive energy of water flowing inland (Kathiresan and Rajendran 2005). Mangrove belts of several hundred meters wide are shown to reduce tsunami height by 5- 30%. Dense and wider mangrove forests are more effective at reducing tsunami height, as well as speed of the water and the area flooded by the tsunami. However, large tsunamis with more than 4 m wave height can damage mangroves and make them less effective at reducing tsunami flows. The protective role of mangroves is location-specific, depending on the tree density and width of forest and the diameter of trunks and roots, along with other factors (e.g., floor shape, bathymetry, and spectral features of waves). So precise site-specific information is imperative for efficient use of mangroves for their protective capacity. In addition, coastal wetlands including mangroves have long been recognised for their capacity to improve water quality by removing excess nitrogen (Land et al. 2016). Three processes that remove nitrogen (N) are plant uptake, soil accretion, and denitrification. Of which, denitrification the conversion of nitrate (NO^{-}_{3}) to nitrous oxide (N_2O) and finally to nitrogen gas (N_2), is the major pathway for permanent N removal from an ecosystem (Kulkarni et al. 2008). Anaerobic ammonium oxidation (anammox) is an additional pathway for N removal through the conversion of nitrite (NO_2) and ammonium (NH^{+}_{4}) to N_2. The wetlands can only improve water quality if they are hydrologically connected, i.e., if they are in contact with water long enough for denitrification to occur. Thus, hydrological connectivity is critical for the provision of ecosystem services through improvement of water quality (Adame et al. 2019).

Livelihood Generation

A wide range of goods and services of offered by mangroves supports the livelihood of the local communities. In Bangladesh and India, around 9 million people are dependent on the Sundarbans for their livelihoods (PDO-ICZMP 2004, GoI 2005). Around 200,000 fishermen are involved in fishing activities in the Sundarbans water. Another 225,000 population, that is 14 % of the people residing within 10 km around the Sundarbans is dependent on the collection of *Penaeus monodon* fry (Hoq 2007). Mangrove woods were used for fuel (firewood or charcoal making) and construction (house, boats and fish stakes). However, considering the threatened status of mangroves, felling of mangrove trees is completely banned in many countries including India. The livelihood activities in mangroves areas can be classified into four broad groups, viz. (I) Agriculture and Aquaculture (II) Fishing and prawn-seed collection, (III) Tourism (IV) Honey and other Non-Timber Forest Products (NTFP) collection (thatch, beverage, sugar, alcohol, wax, tannins and dyes, gums and resins and vinegar). Mangrove species with medicinal properties are also harvested for herbal remedies by coastal communities. Mangrove and Marine Biodiversity Conservation Foundation established by the Government of Maharashtra in September 2015 is implementing sustainable livelihood activities such as mud crab farming, oyster and mussel farming, Asian Seabass cage culture and Ornamental Fish Culture in the mangrove areas of Maharashtra (Vasudevan 2017).

Livelihood activities in mangroves

Mangroves have potential to adapt and mitigate the impacts of climate change processes. The climate change process includes increased CO_2, increased temperature, sea level rise (SLR), changing ocean currents, changes in precipitation, increased natural calamities and Ocean acidification. These changes are likely to have a substantial impact on biodiversity. Climate change processes are highly inter-related. The impacts of climate change are geographically variable at both a regional and local scale in terms of their biodiversity, ecological structure, physiology, hydrodynamic and geomorphological settings, and tolerance to temperature and salinity. These climate change processes can also have both destructive and constructive impacts on mangrove ecosystems.

For instance, the mangroves expand into salt marsh communities as a result of increasing temperatures and a reduction in frost events at higher latitudes in both the northern and southern hemispheres (Cavanaugh et al. 2013; Saintilan et al. 2014). Mangroves are photosynthetically active at or below 30 °C, and their CO_2 assimilation rates decline, either sharply or gradually, as temperature increases from 33 to 35°C (Ball and Sobrado 2002). However, more evidence is needed to show whether mangrove forests in the tropics may experience range contraction in response to increasing temperatures and drought. Mangrove photosynthesis ceases and productivity decreases when leaf temperatures exceed 38–40°C. Furthermore, high temperatures increase evaporation rates, which can result in salinity increases; and the synergistic impacts of salinity and aridity can cause the mortality of mangroves, as witnessed from the massive mangrove dieback in 1000 km stretch of Gulf of Carpentaria, Australia.

Climate change is likely to cause extreme changes in rainfall around the world, with substantial regional variation. Increase in rainfall leads to expansion of mangroves landward (Eslami-Andargoli et al. 2009) as well as seaward (Ashbridge et al. 2016) due to higher supply of fluvial sediments, nutrients, lower exposure to sulphates and reduced salinity. Increase in precipitation is likely to increase riverine discharge, which will increase allochthonous sediment inputs in estuarine mangroves, mangrove surface elevation, and resilience to SLR (Ranasinghe et al. 2013). However, decreases in precipitation and increases in evaporation lead to increase of soil salinity that decreases mangrove productivity.

Mangroves are sensitive to changes in salinity, duration and frequency of inundation; thus, sea level rise is a major threat to mangrove ecosystems. Rise in flooding duration can lead to mangrove plant death at the seaward margins (He et al. 2007) and change in species composition (Gilman et al. 2008), ultimately leading to a reduction in productivity (Castañeda-Moya et al. 2013) and ecosystem services. However, mangroves have the potential to cope up with sea level rise in site-specific manner through vertical accretion and/or plant migration towards inland. Vertical accretion potential of mangroves is influenced by sediment supply, autochthonous peat production, land uplift/subsidence and localized sediment auto-compaction, whereas inland migration depends on availability of accommodation space towards landward. Unfortunately, upstream developmental activity viz., damming/diversion of rivers significantly has reduced the sediment load reaching the coasts and also the coastal developmental activities impede the landward expansion of mangroves. Thus, 69% of Indo-West pacific mangroves are not building surface elevations at the rates that equalled or exceeded sea-level rise and the region remains highly vulnerable (Lovelock et al. 2015).

Mangroves are highly resilient to the damage caused by storms. The frequency and intensity of extreme weather events (typhoons, cyclones, hurricanes) are predicted to be increases in the future. Rapid sediment input and nutrient pulses that can occur during intense storms have the potential to maintain soil elevation in the face of SLR and stimulates productivity and growth. However, physical damage caused by the increasing frequency and intensity of extreme events have the potential to offset the resilience and recovery of mangroves.

The world's oceans absorb increased levels of atmospheric CO_2, and hence the process of hydrolysis increases the potential of hydrogen (pH) of seawater. However, mangroves can counteract the ocean acidification in tropical waters by exporting dissolved inorganic compounds and alkalinity, as proved recently in six pristine mangrove creeks in Australia (Sippo et al. 2016).

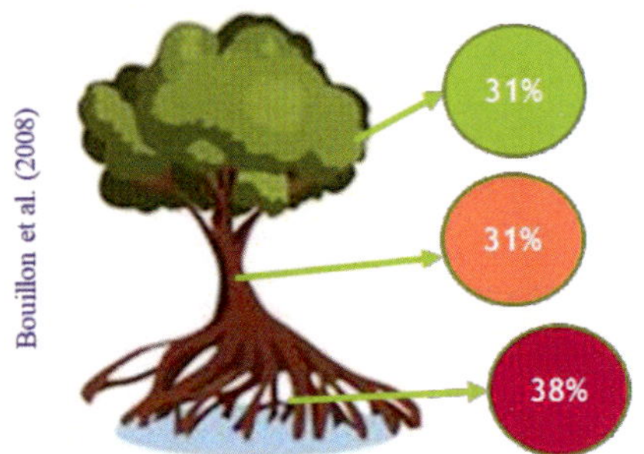

Global Mangrove Carbon Stock

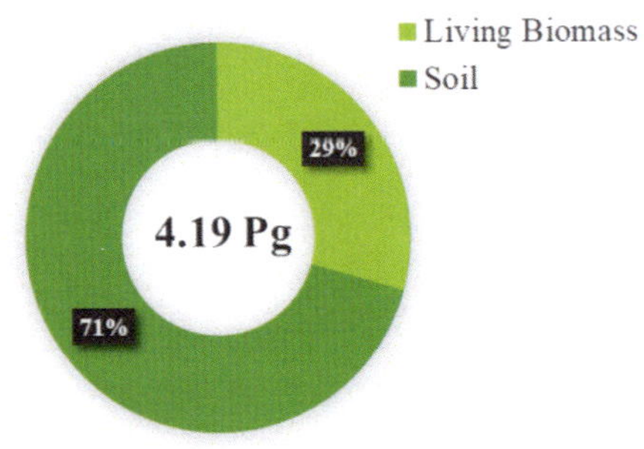

Fate of Mangrove production

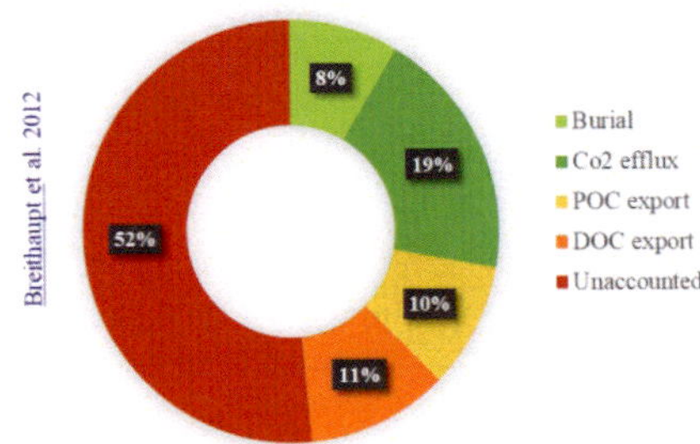

Carbon Storage Potential of Mangroves

Responses of mangroves to increasing atmospheric CO_2 are complex, species specific and confounded by variations in salinity, nutrient availability, and water-use efficiency. Higher CO_2 concentrations can enhance the growth of mangroves with only 7 % increase in net primary productivity, and temperature is an important driver rather than increasing CO_2.

Mangrove forests are the potential carbon sinks for mitigating the growing greenhouse gas emissions due to their highest carbon storage capacity per unit area compared to terrestrial forests. Mangroves are more efficient in carbon storage by 4-fold and sequestration by 10-fold than other tropical forests. Unlike other tropical forests, for which the bulk of carbon storage is in the biomass, mangrove carbon is primarily stored in anaerobic soil for a long period of time. The rough estimate of total mangrove net primary production (NPP) is 209.6 Tg C year^{-1} (Alongi and Mukhopadhyay 2015), of which, ~20% (34.1 Tg C year^{-1}) of the carbon is lost or recycled via CO_2 flux to the atmosphere and ~60% (117.9 Tg C year^{-1}) are exported as particulate organic carbon (POC), dissolved organic carbon (DOC), and dissolved inorganic carbon (DIC) to the ocean (Alongi and Mukhopadhyay 2015; Rosentreter et al. 2018). Of the remaining carbon, burial accounts for 18.4-34.4 Tg C year^{-1} (Breithaupt et al. 2012), and this carbon is considered to be a significant long-term storage of atmospheric CO_2 (Rosentreter et al. 2018). The estimated average soil carbon concentration of global mangroves is 2.6-6.4 Pg C, for the top metre of soil and the estimated of soil carbon stock per unit area of mangroves is 55-1376 Mg C/ha. However, severe lack of country-wise data testifies the role of mangroves in climate change mitigation as most effective at the national level rather than on a global scale. Moreover, carbon storage potential of mangroves is controlled by local biotic and abiotic factors, so it is vital to understand the spatial distribution of mangrove soil carbon stocks across mangrove-lined countries for recognizing the actual climate mitigation potential of this ecosystem so as to strengthen the conservation measures.

Ecosystem	Carbon Stock (Mg C/ha)	Range (Mg C/ha)
Mangroves	**386**	**55-1376**
Tidal Salt marsh	**255**	**16-623**
Seagrasses	**108**	**10-829**

Petal persistence in *Sonneratia alba*

Bioprospectives

Mangroves are widely used in folklore medicine. They are mainly used to treat diabetes, hypertension and gastrointestinal disorders like constipation, diarrhoea, dysentery, dyspepsia, haematuria, and stomach pain. The most common pharmacological activities reported are antioxidant and antimicrobial properties of mangrove extracts. Terpenoids, tannins, steroids, alkaloids, flavonoids, and saponins are the main classes of phytochemicals isolated from mangroves. The most commonly used mangroves are *Bruguiera gymnorhiza* (17%), *Rhizophora mucronata* (14%), *Acanthus ilicifolius* (10%), and *Heritiera fomes* (9%). Mangroves are mostly used in Asian countries, namely India (45.8%), Bangladesh (5.1%), Malaysia (5.1%), China (5.1%), Indonesia (3.4%), Philippines (3.4%) and others (16.9%) (Bibi et al. 2019). India is a pioneer in bioprospecting the mangroves, which are a rich source of salt-resistant genes, novel chemicals and high value products such as (i) black tea beverage, (ii) mosquito repellents, (iii) lignins for controlling oral and cervical cancers, (iv) polysaccharides from preventing the Human Immunodeficiency Virus (HIV) that causes AIDS, (v) anti-diabetic extract, (vi) hair growth stimulant, and (vii) rapid synthesis of nanoparticles (Kathiresan and Ravikumar 2010). Only 26 species are mentioned in literature to possess folklore medicinal importance. But they are largely lacking scientific validation. The endophytic microorganisms of mangroves species possess great potential to offer novel bioactive compounds. Further studies will lead to development of patents, processes, and valuable products.

Calyx of *Sonneratia ovata*

Part 6: Conservation and Management

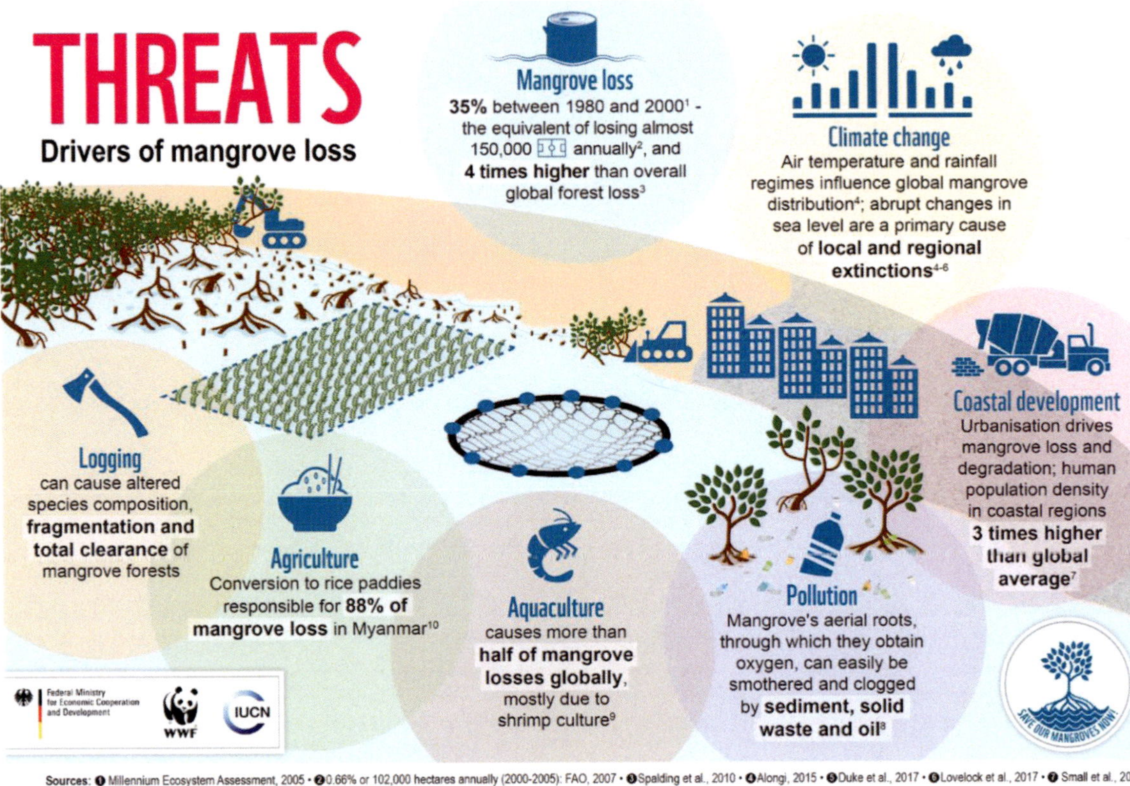

Threats

There are sixteen threat factors have been identified from Indian mangroves (Kathiresan 2010). However, habitat conversion for agriculture and aquaculture activities, environmental pollution by upstream anthropogenic activities, Climate change processes (e.g. sea level rise; increasing frequency of cylones, etc.) are the major threat factors for mangroves. Mangroves are able to cope up with short-term oscillations and long-term fluctuations of climatic conditions, but the consequences of global climate change, such as increased extreme events, will have unprecedented effects on biota and threaten the resilience and recovery potential of the ecosystems (Harris et al. 2018; Sippo et al. 2018). Despite conservation measures, the mangroves still experience an annual loss of 0.2–0.7% during 2000-2012 and they remain the most threatened ecosystem of the world (Hamilton and Casey 2016).

Insect herbivory at Mumbai coast

Legislative measures

Flowers of
Aegiceras corniculatam

India is a good example for conservation and management of mangrove ecosystems (Bhatt and Kathiresan 2012). Felling of mangrove trees is completely banned in India. It is a signatory member of many international conventions pertaining to conservation of species and ecosystems (DasGupta and Saw 2013). India formulated a comprehensive management plan to conserve mangroves. A National Mangrove Committee (NMC) was formed in 1976, as a first step, to advise on conservation measures and to form a framework for effective implementation. Indian mangrove habitats are legally protected under the Indian Forest Act of 1927, various state forest acts, the Forest (Conservation) Act of 1980, the Wildlife (Protection) Act of 1972, the Coastal Regulation Zone (CRZ) Notification (2018) under the Environmental Protection Act of 1986, the Environmental Impact Assessment Notification (EIA) of 1994, and the Coastal Aquaculture Authority Act of 2005. Furthermore, National wetland management Programme (NWMP), National Lake Conservation Plan (NLCP), National River Conservation Plan (NRCP), and Wetlands (Conservation and Management) Rules, 2017 offer legislative protection. In addition, the Coastal Regulation Zone (CRZ) notification (2018) also offers legal protection to mangroves and classified mangroves as ecologically sensitive areas and declared them under the CRZ Category I.

Occurrences pinkish petals in *Sonneratia alba*

Restoration and Rehabilitation

Governments have collaborated with national and international non-governmental organizations (NGOs) to undertake rehabilitation programs in many mangrove degraded areas. For an instance, the M.S. Swaminathan Research Foundation (MSSRF; based in Chennai), in collaboration with the Union Ministry of Environment and Forests and the State Forest Departments, has become actively involved in a mangrove rehabilitation with local community participation since 1996. In high-amplitude areas, the existing planting techniques of direct seed sowing and planting of seedlings in mud flats are followed, whereas in low-amplitude areas (in Tamil Nadu and Andhra Pradesh), canal bank planting is used. This technique was first attempted in 1987 in mangrove forests at Muthupet, Tamil Nadu, and different models have been developed (Baruah 2004). A “fish bone” design is the most successful one for canal bank planting. This is also replicated in other mangrove forests.

Sonneratia alba at sandy coast of Little Andaman

Success Stories

Mangroves are well conserved and managed in India. This is evident by the increasing mangrove forest cover, in every two years since 1995 as reported by Forest Survey of India, Dehradun. The mangrove forest cover increased by 112, 181 and 54 km^2 respectively during 2013-2015, 2015-2017 and 2017-2019. However, India had a mangrove cover of 6000 km^2 during year the 1960s, and it has reduced by 17%, i.e. 4975 km^2 in 2019. However, since 1995, the mangrove cover has got stabilized close to 4500 km^2 with an increasing trend, despite increasing pressures. However, it is necessary to achieve a target of 6000 km^2 by restoring the mangroves in potential areas within a period of 10 years (Kathiresan 2018). Most of the mangrove restoration programs in India have intended to increase the area of coverage, and most afforestation efforts have used a few fast-growing species of Avicennia and Rhizophora, with a mixed survival rate. For instance, Sanyal et al. (1998) reported that between 1989 and 1995, 9050 ha of mangroves was planted in West Bengal but with only a 1.52% success rate, whereas the survival rates have been high in Andhra Pradesh (55–93%), Tamil Nadu (80–100%), Karnataka (45–95%), and Kerala (52–85%).

Mangrove restoration has been successfully demonstrated. From 2002 to 2006, 41.95 km2 of mangrove areas in Southern India were restored (Sahu et al. 2015; Bhatt and Kathiresan 2011). The largest area of mangroves restored is in Andhra Pradesh (19.78 km^2 since 1987), mostly in the East Godavari and Krishna districts. In Tamil Nadu, about 8.40 km^2 of mangroves (3.45 km^2 at Pichavaram, 2.95 km^2 at Muthupet, and 2 km^2 at Ramanathapuram) were restored through plantation (Bhatt and Kathiresan 2011). In Kerala, mangrove restored is 1.34 km^2 in the Kannur Division since 1997. In Karnataka, the total mangrove area restored since 2000 is 12.44 km^2. In Gujarat, 384.08 km^2 of mangroves were established during 2001–2009 by the Gujarat Forest Department. In addition, the Gujarat Ecology Commission, Gandhinagar, increased the mangrove area of 55.46 km^2 during 2001–2007 by involving local communities (Pandey and Pandey 2011). The efforts are continued in the country with the financial support of central and state governments. India has demonstrated the best practices of conservation and management of mangroves, and they are (i) Canal bank planting with 'Fish Bone' design for mangrove restoration; (ii) Maharashtra Mangrove Conservation Model; (iii) Kannur Mangrove Mission and (iv) Participatory mangrove management model.

Canal Bank planting with 'Fish Bone' design

Mangroves are largely degraded in Tamil Nadu and Andhra Pradesh. The reason for the degradation is attributed to high salinity of dry soil as a result of lack of regular tidal flushing due to low tidal amplitude (Kathiresan 2000). To overcome this situation, the M. S. Swaminathan Research Foundation (Chennai) and the Forest Department demonstrated the 'Canal-Bank Planting' technique with 'fish bone' design, in Muthupet and Pichavaram in Tamil Nadu as well as East Godavari and Krishna districts of Andhra Pradesh (Baruah 2012, Ramasubramanian and Ravishankar 2004). In this technique, canals are formed so that the high saline soil gets regular tidal inundation, leaches out salts and becomes suitable for mangrove restoration. This effort was undertaken with the participation of local mangrove user communities resulting in increased forest cover by about 90% in the degraded areas of Pichavaram mangrove wetland, between 1986 and 2002 as proved by satellite data (Selvam et al. 2003).

Hydrological restoration implemented using canal bank planting in mangrove wetlands in Pichavaram

Canal bank planting at Pichavaram

Rhizophora mucronata at Cochin Coast

Maharashtra Mangrove Conservation Model

The Government of Maharashtra initiated mangrove conservation, after the landmark order of the Bombay High Court on 6 October 2005. The high court order prohibited all constructions in mangrove areas as well as within 50 m radius from the mangrove boundary. It also directed that mangroves on government land be declared as 'protected areas' under the Indian Forest Act and transfer them to the forest department. As a result, 5469 ha of mangroves on government land were transferred to the Thane Forest Division by the revenue authorities. (Vasudevan 2017). The Government of Maharashtra constituted 'Mangrove Cell' in January 2012, and then the 'Mumbai Mangrove Conservation Unit' in 2013 to protect mangroves in Mumbai and the adjoining areas, and elevated the status of mangrove forests on government land from 'protected forests' to 'reserved forests'. This led to notification of 15,088 hectares of mangroves on Government land as 'reserved forests' in seven districts of Maharashtra. The mangrove cell also achieved plantation in more than 200 ha of mangrove degraded areas in the Greater Mumbai region. In addition, the cell demarcated the areas under the control of Forest Department on the ground with boundary markings, based on satellite mapping at 1:50,000 scale. As a result of all these efforts, the mangrove cover has significantly increased since 2013(Vasudevan 2017; FSI 2019). In addition Mangrove Cell and Mangrove foundation implementing sustainable aquaculture livelihood activities in more than hundred villages along the Coastline of Maharashtra . The aquaculture activities being implemented are mangrove mud crab farming , bivalve farming (Oyster and Mussel), Asian seabass cage culture, and ornamental fish culture.

Kannur Mangrove Mission and Participatory mangrove management

Rhizophora spp., at Pichavaram

The 'Mission Mangrove Kannur' was taken up by the District Collector, Kannur (Kerala), along with the forest department in May 2014 to survey, notify and save mangroves of the district. The mission took 14 months to achieve its goal for the first time in Kerala. As a result, a comprehensive survey of mangroves was completed and 236 ha of mangroves were notified as 'Reserved Forest', for permanent conservation. In addition, the mission started the process of acquisition of 1,200 acres of mangroves from private owners (Bala Kiran 2017).

In India, mangroves are managed prominently in the states of Tamil Nadu, Odisha, Andhra Pradesh, West Bengal and Gujarat through community-based co-management (Kathiresan 2005). With financial support from Indo-Canada Environmental Facility (ICEF), New Delhi, the M. S. Swaminathan Research Foundation achieved Joint Mangrove Management (JMM) for restoration and conservation of mangroves through the participation of local people along with the forest departments. The JMM project involved 5240 families from 28 villages along the east coast of India. About 1475 ha of mangroves were restored by planting 6.8 million mangrove saplings. To empower local people, 194 self-help groups were organized to implement poverty alleviation programmes such as supplementary income-generating activities for firewood, fodder, fencing and house construction. A similar effort was undertaken by the Gujarat Ecology Commission with financial support of ICEF, New Delhi. This project has promoted community-based regeneration and management of the mangroves in about 5000 ha area along the Gulf of Kachchh and Gulf of Khambhat in five years from 2001 to 2006.

Fruit of *Nypa fruticans*

Rhizophora spp., at sandy beach of Havelock Island

Effectiveness of the Existing Conservation measures

Four activities are suggested for improving mangrove management in several countries, and they are (1) acquisition of private lands with mangroves, (2) legal protection of mangroves, and (3) Declaration of protected areas and (4) massive mangrove rehabilitation programs (Lewis et al. 2016). Of these, the first two are either not strictly enforced or remaining as paper policy in many countries (Amir 2018; Castellanos-Galindo et al. 2017). Worldwide there are some 2,500 protected areas that include mangrove forests within their boundaries, which represents over 39% of world's remaining mangroves (IUCN and UNEP-WCMC 2019). However, many of these PAs are poorly designed or poorly enforced due to inadequate manpower, lack of facilities and absence of formal management plans (Singh 2003; Lavieren et al. 2012), and some PAs fail to prevent mangrove loss and degradation within their ranges (Lavieren et al. 2012; Almeida et al. 2016). The effectiveness of PAs with respect to ecological and/or socio-economic factors is debatable (Bennett and Dearden 2014), and most of the PAs lack both clear aims and long-term biological data to evaluate the management effectiveness (Addison et al. 2015). Furthermore, ecosystem connectivity and climate change considerations are often lacking in the PAs of coastal and marine environments (Mcleod et al. 2009; Magris et al. 2014). Protected areas prevent some drivers of degradation, such as unsustainable timber extraction, and allow recovery. However, other drivers of degradation, such as upstream water abstraction or changes to sediment supplies, cannot be influenced when they occur beyond the protected area boundaries (Worthington and Spalding 2019). Recently it has be estimated that rates of degradation of mangroves in protected areas is 0.57% and about 6% of mangrove areas within protected areas have been lost since 1996 (Worthington and Spalding 2019). Despite of existing guidance for successful rehabilitation effort (Lewis and Brown 2014), most of the mangrove planting efforts are not largely successful (Bosire et al. 2008; Dale et al. 2014; Brown et al. 2014; Begam et al. 2017). Rehabilitation/Restoration of degraded mangroves is often based on mono-specific plantation without understanding the species-specific information or habitat characteristics. Thus, despite the short-term increases in area, most of the mangrove restoration/rehabilitation efforts could not offset for the continual destruction of forests. Unsuccessful rehabilitation is a waste of time, money and human resources. Further, the habitat characteristics, altered for the purpose of mangrove rehabilitation render these corridors unsuitable for natural migration of mangroves to cope up with the impacts of climate change. Large-scale planting, where survival is low or, worse, is likely to result in collateral damage to existing or adjacent habitats (Lee et al. 2019). Hence, another approach, called "Early detection and pre-emptive rehabilitation" is suggested by Lewis et al. (2016) to prevent complete loss of plant community structure and ecological function through long-term monitoring of changes in hydrological and ecological status of mangroves. However, such long-term monitoring efforts is not implemented in any country (Duke et al. 2017).

Enhancing the adaptive potential of mangroves - Need of the hour

Mangrove degradation prevails globally as well as in India. For instance, earthquake induced geomorphological changes in A & N Islands resulted in the degradation of 65km^2 mangrove area in North Andaman Island (Ramakrishnan et al. 2020) and 97% of mangroves in Nicobar Islands (Nehru and Balasubramanium 2018). Moreover, ecological health of the Indian mangroves is declining due to anthropogenic pressure (Kathiresan 2018). So, it is imperative to adjust the existing conservation measures to improve the ecological heath of exiting mangroves rather than increasing extent of their area alone. It is widely recognized that adaptive potential promotes resilience of species to recover from disturbances, and thereby improving the effectiveness of conservation practices (Intergovernmental Panel on Climate Change (IPCC) 2014). Thus, determining the physiological, ecological and genetic processes underpinning the adaptive potential of species is a key focus for conservation approach (Osland et al. 2019; Rilov et al. 2019; Wee et al. 2019). Adaptive potential is often defined as the ability of populations or species to adapt to rapid environmental change with minimal disruption by means of phenotypic or molecular changes (Eizaguirre and Baltazar-Soares 2014). Adaptive potential of populations/species often measured in terms of genetic diversity. Thus, use of genetic information in conservation is crucial for its long-term effectiveness to preserve the adaptive and evolutionary potential of ecosystem/species (Hoffmann and Sgro 2011). However, little has been translated into on-the-ground conservation (Wee et al. 2019). Despite being threatened by human developmental activities and climate change, the mangrove forests are highly resilient ecosystems that have the potential to adapt and adjust to changing conditions (Woodroffe et al. 2016). This is evident by recent expansion of mangroves towards polar regions in response to increasing minimum winter temperature (Saintilan et al. 2014) as well expansion and contraction in response to temperature fluctuations and Pleistocene sea level drop and rise in the long past (Ludt and Rocha 2014). Vertical adjustment and horizontal movement across the landscape are the processes that govern the responses of mangrove forest to sea-level rise (Woodroffe et al. 2016; Krauss et al. 2014; Lovelock et al. 2015). Furthermore, the observed higher expression of biodiversity than the genetic diversity and smaller genome size warrant the significant evolutionary potential of mangroves (Lyu et al. 2018; Lira-Medeiros et al. 2010; Wee et al. 2018). Hence, maximizing the adaptive potential of mangroves is need of the hour to safeguard the existing mangroves and sustain the provision and use of their manifold services (Osland et al. 2018) rather than short-term increase in mangrove cover by unproductive restoration/rehabilitation efforts (Lee et al. 2019).

Growth of mangroves on newly formed intertidal areas of Great Nicobar Island (Galathea Bay)

Strategies Needed

Mangrove biodiversity, structure and processes are site-specific (Duke et al. 1998; Binks et al. 2018). Within a given mangrove habitat, local abiotic and biotic factors (e.g., hydrology, geomorphology, salinity, competition, facilitation etc.,) greatly influence the mangrove diversity, structure and processes, despite the climatic drivers (i.e., temperature and precipitation regimes) that are controlling mangrove distribution and species richness on global scale. Furthermore, mangroves are highly dynamic and have strong interconnection with adjacent habitats. Thus, the potential strategies to enhance the adaptive potential of mangroves are facilitating the connectivity between adjacent ecosystems, ensuring the minimum hydrological changes, and protection of natural corridors. This can be achieved through "Ecosystem based management (EBM)" by making use of site-specific (biodiversity, habitat characteristics, ecological process) and species-specific information (distribution, habitat requirements). EBM is driven from the failure of conventional management practices to protect marine ecosystems from overexploitation (Crain et al. 2009). EBM aims at achieving conservation, sustainable use and fair allocation of benefits from natural resources, thereby striking a balance between short-term needs and sustainability (Cowan et al. 2012). EBM requires more precise information of various components of a given ecosystem to be managed. Thus, it is still a long way to achieve the formulation of EBM for coastal wetlands, but site-specific EBM can be initiated with available information and to undertake parallel efforts fulfilling the needs of effective implementation. So, it is necessary to critically revisit the available information by multi- and interdisciplinary teams and to make efforts in filling the knowledge gaps in site specific and species-specific information for better policymaking.

Bruguiera gymnorhiza

Future directions

- Long term monitoring studies on changes in coverage and composition of foundation species of coastal wetlands and dynamics of coastal geomorphology
- Estimation of coastal wetland habitat loss or being lost or Linkages/connectivity among coastal habitats/ecosystems
- Study of vital functional processes in coastal wetland ecosystems
- Study of structurally important elements of coastal wetland ecosystems
- Ecological services of coastal wetlands and their economic valuation
- Site-specific understanding of natural process (viz. vertical accretion) in coastal wetlands to cope up with climate change events like sea-level rise
- Monitoring the effect of anthropogenic disturbance (i.e. eutrophication) in coastal aquatic environments
- Rehabilitation of degraded mangrove areas by standard guidelines
- Development and diversification of mangrove biodiversity in restoration programs
- Recovery of mangrove species under threat
- Focus on participatory approach for mangrove conservation and management
- Studies on pollinators and improving strategies for mangrove reproductive potential
- Creation of a Gene Bank for coastal flora
- Invasive species and their threats
- Review and Management Effectiveness of PAs in coastal marine environments

Fruits of *Cynometra iripa*

References

- Adame MF, Franklin H, Waltham NJ, Rodriguez S, Kavehei E, Turschwell MP, et al. (2019) Nitrogen removal by tropical floodplain wetlands through denitrification. Mar. Freshw. Res. 70:1513–1521. doi: 10.1071/MF18490.
- Addison PFE, Flander LB, Cook CN (2015) Are we missing the boat? Current uses of long-term biological monitoring data in the evaluation and management of marine protected areas. Journal of Environmental Management 149:148–156.
- Almeida LTD, Olimpio JLS, Pantalena AF, Almeida BSD, Soares MDO (2016) Evaluating ten years of management effectiveness in a mangrove protected area. Ocean & Coastal Management 125:29–37.
- Alongi DM (1988) Bacterial productivity and microbial biomass in tropical mangrove sediments. Microb Ecol 15:59–79.
- Alongi DM, Mukhopadhyay SK (2015) Contribution of mangroves to coastal carbon cycling in low latitude seas. Agric For Meteorol 213:266–272.
- Amir AA (2018) Mitigate risk for Malaysia's mangroves. Science 359(6382): 1342-1343.
- Anant PT (2012) Study of mangrove flora along the Zuari river (Case study on curtorim village-Goa- india). International Research Journal of Environmental sciences. 1(5):35-39.
- Anupama C, Sivadasan M (2004) Mangroves of Kerala, India. Rheedea 14:9–46.
- Arisdason W, Magesh CR, Dinesh Albertson W, Venu P (2008) Mangroves of Andhra Pradesh: Some Issues in Taxonomy, Rarity and Conservation. Proc. A. P. Akad. Sci. 12: 193-202.
- Arora RK, Agarwal KR (1960) Observation on the vegetation on the Malpe Coast and neighbouring Islands. J. Indian Bot. Soc. 44: 314-324.
- Arunprasath A, Gomathinayagam M (2014) Distribution and Composition of True Mangroves Species in three major Coastal Regions of Tamil Nadu, India. International Journal of Advanced Research. 2: 241-247.
- Bala Kiran P (2017) Mission Mangroves Kannur, Mangal-Van, Mangrove Society of India, Goa, pp. 41–54.
- Baruah AD (2004) Muthupet mangroves and canal bank planting technique. Nagapattinam.
- Baruah AD (2012) Muthupet mangroves and canal bank planting technique. Tamil Nadu Forest Department, p. 61.
- Basha CS (1991) Distribution of mangroves in Kerala. Indian Forester 117:439–449.
- Basha CS (1992) Mangroves of Kerala: a fast disappearing asset. Indian Forester 118(3):175–190.
- Begam MM, Sutradhar T, Chowdhury R, Mukherjee C, Basak SK, Ray K (2017) Native salt-tolerant grass species for habitat restoration, their acclimation and contribution to improving edaphic conditions: a study from a degraded mangrove in the Indian Sundarbans. Hydrobiologia. doi:10.1007/s10750-017-3320-2.
- Bennett NA, Dearden P (2014) From measuring outcomes to providing inputs: governance, management, and local development for more effective marine protected areas. Marine Policy 50:96–110.
- Bernini E, Rezende GE (2011) Vegetation structure in a mangrove forest in southeastern Brazil. Pan-Am J Aquat Sci 6(3):193–209.
- Bharucha FR, Navalkar BS (1950) Studies in Ecology of Mangroves. J. Univ. Bombay 18: 7-16.
- Bhatt JR, Kathiresan K (2011) Biodiversity of Mangrove Ecosystems in India. In Towards Conservation and Management of Mangrove Ecosystems in India/ed. by Bhatt J.R., D.J. Macintosh, T.S. Nayar, C.N. Pandey and B. P. Nilaratna, IUCN India. pp 1-34.
- Bhatt JR, Kathiresan K (2012) Valuation, carbon sequestration potential and restoration of mangrove ecosystems in India; In Sharing Lessons on Mangrove Restoration Proceedings and a Call for Action from an MFF Regional Colloquium 30–31 Aug. 2012, Mamallapuram, India. 19–38pp.
- Bhatt S, Shah DG, Desai N (2009) The mangrove diversity of Purna estuary, South Gujarat, India. Tropical Ecology. 50: 287-293.
- Bhosale LJ, Banik S, Gokhale MV, Jayappa MA (2002) Occurrence of Xylocarpus granatum Koen and Cynometra iripa Kostel along the coast of Maharashtra, J. Econ. Taxon. Bot. 26(1): 82 – 87.
- Bhosale LJ, Banik S, Gokhale MV, Jayappa MA (2002) Occurrence of *Xylocarpus granatum* Koen and *Cynometra iripa* Kostel along the coast of Maharashtra. J. Econ. Taxon. Bot. 26: 82 – 87.
- Bibi NS, et al. (2019) Ethnopharmacology, Phytochemistry, and Global Distribution of Mangroves—A Comprehensive Review. Marine Drugs 17: 231. doi:10.3390/md17040231.
- Binks RM, Byrne M, McMahon K, Pitt G, Murray K, Evans RD (2018) Habitat discontinuities form strong barriers to gene flow among mangrove populations, despite the capacity for long-distance dispersal. Diversity and Distributions. doi:10.1111/ddi.12851.
- Blasco (1977) Outline of ecology, botany and forestry of mangals of the Indian subcontinent. In: Chapman VJ (ed) Wet coastal ecosystem. Elsevier, Amsterdam.
- Blasco F (1975) The Mangroves of India. French Institute, Pondicherry. Trav SCI Tech 4:163.
- Blatter E (1905) The mangroves of Bombay Presidency and its biology. J. Bombay Nat. Hist. Soc. 16: 644-656.
- Bosire JO, Dahdouh-Guebas F, Walton M, Crona BI, Lewis RR, Field C, Koedam N (2008) Functionality of restored mangroves: a review. Aquatic Botany 89(2): 251–259.
- Breithaupt JL, Smoak JM, Smith TJ, Sanders CJ, Hoare A (2012) Organic carbon burial rates in mangrove sediments: strengthening the global budget. Glob iogeochem Cycle 26(3):GB3011.

Fertile leaflets of *Acrostichum speciosum*

References

- Breteler FJ (1977) America's Pacific species of Rhizophora. Acta Bot Neerl 26:225–230.
- Brown B, Fadilla R, Nurdin Y, Soulsby I, Ahmad R (2014) Community based ecological mangrove rehabilitation (CBEMR) in Indonesia. SAPIENS 7: 53–64.
- Bunt JS (1996) Mangrove zonation: an examination of data from seventeen riverine estuaries in tropical Australia. Ann Bot 78:333–341.
- Bunt JS (1999) Overlap in mangrove species zonal patterns: some methods of analysis. Mangrove Salt Marshes 3:155–164.
- Bunt JS, Bunt ED (1999) Complexity and variety of zonal pattern in the mangroves of the Hinchinbrook area, Northeastern Australia. Mangrove Salt Marshes 3:165–176.
- Bunt JS, Stieglitz T (1999) Indicators of mangrove zonality: the Normanby river, NE Australia. Mangrove Salt Marshes 3:177–184.
- Bunt JS, Williams WT, Hunter JF, Clay HJ (1991) Mangrove sequencing: analysis of zonation in a complete river system. Mar Ecol Prog Ser 72:289–294.
- Castellanos-Galindo GA, Klunger LC, Tompkins P (2017) Panama's impotent mangrove laws. Science 355 (6328): 918-919.
- Cerón-Souza I, Gonzalez EG, Schwarzbach AE, Salas-Leiva DE, Rivera-Ocasio E, Toro-Perea N, Bermingham E et al (2015) Contrasting demographic history and gene flow patterns of two mangrove species on either side of the Central American isthmus. Ecol Evol 5:3486–3499.
- Chapman VJ (1975) Mangrove biogeography. In: Walsh GE, Snedaker SC, Teas H (eds) Proceedings of international symposium on biology and management of mangroves. Institute of Food and Agricultural Sciences/University of Florida, Gainesville, pp 3–22.
- Chavan NS (2013) New area record of some mangrove species and associates from the coast of Maharashtra, India. Seshaiyana 21:2–3.
- Cooke T (1901-1908) The Flora of the Presidency of Bombay. Taylor & Francis, London.
- Costanza R, et al. (1997) The value of the world's ecosystem services and natural capital. Nature 387: 253–260.
- Cowan JH, Rice JC, Walters CJ, Hilborn R, Essington TE, Day JW, Boswell KM (2012) Challenges for implementing an ecosystem approach to fisheries management. Marine and Coastal Fisheries 4:496–510.
- Crain CM, Halpern BS, Beck MW, Kappel CV (2009) Understanding and managing human threats to the coastal marine environment. Conservation Biology 1162:39–62.
- Dagar HS, Dagar JC (1991) Plant folk medicines among the Nicobarese of Katchal Island India. Econ Bot 45:114–119.
- Dagar JC (1982) Some ecological aspects of mangrove vegetation of Andaman & Nicobar Islands. Sylvatrop Philipp Forest Res J 7:177–216.
- Dagar JC, Dagar HS (1986) Mangroves and some coastal plants in ethnobotany of the tribals of Andaman and Nicobar Islands. J Andaman Sci Assoc 2(2):33–36.
- Dagar JC, Dagar HS (1999) Ethnobotany of aborigines of Andaman–Nicobar Islands. Surya International Publications, Dehra Dun, India. pp.203.
- Dagar JC, Mongia AD, Bandyopadhyay AK (1991) Mangroves of Andaman & Nicobar Islands. Oxford/IBH Publishing Co., New Delhi, p 166.
- Dagar JC, Singh NT (1999) Plant resources of Andaman & Nicobar Islands. Bishen Singh Mahendra Pal Singh, Dehra Dun, p 985.
- Dagar JC, Singh NT (1999) Plant resources of the Andaman and Nicobar Islands.Introduction, General Features, Vegetation and Floristic Elements Vo. 1 and 2. Bishen Singh Mahendra Pal Singh, Dehra Dun.
- Dale PER, Knight JM, Dwyer PG (2014) Mangrove rehabilitation: a review focusing on ecological and institutional issues. Wetland Ecology and Management 22: 587–604.
- Dam Roy S, Krishnan P, George G, Kaliyamoorthy M, Goutham Bharathi MP (2009) Mangroves of Andaman and Nicobar Islands, A field guide. Central Agricultural Research Institute, Port Blair, India.
- Das Gupta R, Shaw R (2013) Changing perspectives of mangrove management in India—an analytical overview. Ocean Coast Manage 80:107–118.
- Debnath HS (2004) Mangroves of Andaman and Nicobar Islands; Taxonomy and Ecology A Community Profile. Bishen Singh Mahendra Pal Singh, Dehradun.
- Debnath HSB, Singh K, Giri P (2013) A new mangrove species of *Acanthus* l. (Acanthaceae) from the Sundarbans (India). Indian J For 36:411–412.
- Dodd RS and Rafii ZA (2002) Evolutionary genetics of mangroves: continental drift to recent climate change. Trees 16: 80–86.
- Dodd RS, Afzal-Rafii Z, Kashani N, Budrick J (2002) Land barriers and open oceans: effects on gene diversity and population structure in Avicennia germinans L. (Avicenniaceae). Mol Ecol 11:1327–1338.
- Drury H (1864) Handbook of Indian flora: Being a guide to all flowering plants, vol 1. Travancore Sircar Press, Trivandrum, p 175.
- Duke NC (1991) A systematic revision of the mangrove genus *Avicennia* (Avicenniaceae) in Australasia. Australian Systematic Botany 4: 299–324.
- Duke NC (1992) Mangrove floristics and biogeography. In Tropical Mangrove Ecosystems; Wiley: Hoboken, NJ, USA.
- Duke NC (2006) Australia's mangroves: the authoritative guide to Australia's mangrove plants. The University of Queensland and Norman C. Duke, Brisbane.
- Duke NC (2017) Mangrove floristics and biogeography revisited: further deductions from biodiversity hot spots, ancestral discontinuities, and common evolutionary processes. In: Rivera- Monroy VH, Lee SY, Kristensen E, Twilley RR (e) (eds) Mangrove ecosystems: a global biogeographic perspective. Springer, Cham, pp 17–53.
- Duke NC, Ball MC, Ellison JC (1998) Factors influencing biodiversity and distributional gradients in mangroves. Glob Ecol Biogeogr 7(1):27–47.
- Duke NC, Jackes BR (1987) A systematic revision of the mangrove genus *Sonneratia* (Sonneratiaceae) in Australasia. Blumea 32: 277–302.
- Duke NC, Kovacs JM, Griffith A, Preece L, Hill DJ, van Oosterzee P, Mackenzie J, Morning, HS, Burrows D (2017) Large-scale dieback of mangroves in Australia's Gulf of Carpentaria: a severe ecosystem response, coincidental with an unusually extreme weather event. Marine and Freshwater Research. doi:10.1071/MF16322.
- Dwivedi SN, Parulekar AH, Goswami SC (1974) Ecology of Mangrove Swamps of the Mandovi Estuary, Goa, India. Proc. Inter. Symp. Manag. Mangroves, Hawaii 1: 115-194.

Male Inflorescence of *Excoecaria indica*

References

- Eizaguirre C and Baltazar-Soares M (2014) Evolutionary conservation—evaluating the adaptive potential of species. Evolutionary Applications 7: 963–967.
- Ellison AM (2008) Managing mangroves with benthic biodiversity in mind: moving beyond roving banditry. J. Sea Res., 59:2–15.
- Ellison AM, Farnsworth EJ, Merkt RE (1999) Origins of mangrove ecosystems and the mangrove biodiversity anomaly. Glob Ecol Biogeogr 8(2):95–115.
- Ellison AM, Mukherjee BB, Karim A (2001) Testing patterns of zonation in mangroves: scale dependence and environmental correlates in the Sundarbans of Bangladesh. J Ecol 88:813–824.
- Erlanson EW (1936) A preliminary survey of marine boring organisms in Cochin harbour. Curr Sci 4:726–752.
- Field CD (1995) Impacts of expected climate change on mangroves. Hydrobiologia 295(1–3):75–81.
- FSI (2017) India state of forest report. Forest survey of India, Dehradun, pp 55–61.
- FSI (2019) India state of forest report. Forest survey of India, Dehradun, pp 53–63.
- GEER. 2000. Mangroves in Gujarat. GEER Foundation, Gandhinagar.
- Giesen W, Wulffraat S, Zieren M, Scholten L (2006) Mangrove guidebook for Southeast Asia. FAO and Wetlands International, RAP Publication, Thailand, 769 pp.
- Giri C, Ochieng E, Tieszen L, Zhu Z, Singh A, Loveland T, Masek J, Duke N (2011) Status and distribution of mangrove forests of the world using earth observation satellite data. Glob Ecol Biogeogr 20:154–159.
- Gopal B, Chauhan M (2006) Biodiversity and its conservation in the Sundarban mangrove ecosystem. Aquat. Sci. 68: 338–354.
- Goutham-Bharathi MP, Kaliyamoorthy M, Dam Roy S, Krishnan P, George G, Murugan C (2012) *Sonneratia ovata* (Sonneratiaceae) – A New Distributional Record for India from Andaman and Nicobar Islands. Taiwania. 57: 406–409.
- Government of India [GOI] (2005) Ramsar COP9 national planning tool-national report format India. Source: www.ramsar.org/cop9.
- Govinda-Menon K (1930) Indian medicinal plants. Ramanuja Press, Thrissur.
- Guo Z, Li X, He Z et al (2018) Extremely low genetic diversity across mangrove taxa reflects past sea level changes and hints at poor future responses. Glob Change Biol 24:1741–1748.
- Hamilton SE, Casey D (2016) Creation of a high spatio-temporal resolution global database of continuous mangrove forest covers for the 21st century (CGMFC-21). Global Ecology and Biogeography 25(6): 729–738.
- Harris RMB, Beaumont LJ, Vance TR, Tozer CR, et al. (2018) Biological responses to the press and pulse of climate trends and extreme events. Nature Climate Change 8: 579–587.
- He Z, Li X, Yang M, et al. (2019) Speciation with gene flow via cycles of isolation and migration: insights from multiple mangrove taxa. Natl Sci Rev. 6(2): 275–288.
- Hoffmann AA and Sgro CM (2011) Climate change and evolutionary adaptation. Nature 470:479–485.
- Hoq ME (2007) An analysis of fisheries exploitation and management practices in Sundarbans mangrove ecosystem, Bangladesh. Ocean & Coastal Management 50 (5-6): 411-427.
- Hou D (1960) A review of the genus Rhizophora with special reference to the Pacific species. Blumea 10:625–634.
- IPCC (2014). Climate Change 2014: Mitigation of Climate Change. Contribution of Working Group III to the Fifth Assessment Report of the Intergovernmental Panel on Climate Change. 1454 pp.
- IUCN and UNEP-WCMC (2019): www.protectedplanet.net.
- Jagtap TG (1985) Ecological Studies in Relation to the Mangrove Environment along the Goa Coast, India; a Ph. D. Thesis Submitted to Shivaji University, Kolhapur.
- Jagtap TG, Untawale AG, Inamdar SN (1994) Study of Mangrove Environment of Maharashtra Coast using Remote Sensing Data. Indian Journal of Marine Sciences 23: 90-93.
- Janzen DH (1985) Mangroves: where's the understory? J Trop Ecol 1:89–92.
- Jayatissa LP, Dahdouh-Guebas F, Koedam N. 2002. A review of the floral composition and distribution of mangroves in Sri Lanka. Bot J Linn Soc 138: 29-43.
- Kangas P (2006) Mangrove forest structure on the Sittee River, Belize. Unpublished manuscript, Natural resources management program, University of Maryland, College Park, MD.
- Kathiresan K (1995) *Rhizophora ×annamalayana*: a new species of mangroves. Environment and Ecology 13: 240-241.
- Kathiresan K (2000) Why are mangroves degrading? Curr. Sci., 83: 1246–1249.
- Kathiresan K (2005) Book review: atlas of mangrove wetlands of India. Curr. Sci., 88: 182–183.
- Kathiresan K (2008) Biodiversity of mangrove ecosystems. Proceedings of Mangrove Workshop, GEER Foundation, Gujarat, India, 17pp.
- Kathiresan K (2010) Importance of mangrove forest of India. J. Cost. Environ. 1: 11–26.
- Kathiresan K (2018) Mangrove forests of India. Curr Sci 114(5):976–981.
- Kathiresan K, Anburaj R, Gomathi V, Saravanakumar, K (2013) Carbon sequestration potential of *Rhizophora mucronata* and *Avicennia marina* as influenced by age, season, growth and sediment characteristics in southeast coast of India. J. Coastal Conserv., 17: 397–408.
- Kathiresan K, Bingham BL (2001) Biology of mangroves and mangrove ecosystems. Adv Mar Biol 40:81–251.

Fruits of *Nypa fruticans*

References

- Kathiresan K, Gomathi V, Anburaj R, Saravanakumar K (2014) Impact of mangrove vegetation on seasonal carbon burial and other sediment characteristics in the Vellar-Coleroon estuary. India. J. For. Res., 25: 787–794.
- Kathiresan K, Qasim SZ (2005) Biodiversity of Mangrove Ecosystems, Hindustan Publishing Corporation, New Delhi, India, p. 251.
- Kathiresan K, Rajendran N (2005) Coastal mangrove forests mitigated Tsunami. Estuarine Coastal Shelf Sci., 65: 601–606.
- Kathiresan K, Ravikumar S (2010) Marine pharmacology: an overview. Mar. Pharmacol., 1: 1–37.
- Kathiresan K, Veerappan N, Balasubramanian R (2015) Status of fauna in mangrove ecosystems of India. In: Marine faunal diversity in India, pp 485–497.
- Kim K, Kim H, Lim JH, Lee SJ (2016b) Development of a Desalination Membrane Bioinspired by Mangrove Roots for Spontaneous Filtration of Sodium Ions. ACS Nano 10: 11428–11433.
- Kim K, Seo E, Chang SK, Park TJ, Lee SJ (2016a) Novel water filtration of saline water in the outermost layer of mangrove roots. Scientific Reports 6: 20426. doi: 10.1038/srep20426.
- Komiyama A, Ong JE, Poungparn S (2008) Allometry, biomass and productivity of mangrove forests: a review. Aquat Bot 89:128–137.
- Kothari and Singh (1998). Mangrove Diversity along the North West coast of India. J. Econ. Taxon. Bot. 22: 51-585.
- Kothari MJ, Rao KM (2002) Mangroves of Goa. Botanical Survey of India, Kolkata.
- Kothari MJ, Rao KM (2002) Mangroves of Goa. Botanical Survey of India, Kolkata.
- Krauss KW, McKee KL, Lovelock CE, Cahoon DR, Saintilan N, Reef R, Chen L (2014) How mangrove forests adjust to rising sea level. New Phytol. 202: 19–34.
- Kulkarni MV, Groffman PM, Yavitt JB (2008) Solving the global nitrogen problem: it's a gas! Frontiers in Ecology and the Environment 6: 199–206.
- Kurian CV (1980) Fauna of the mangrove swamps on Cochin estuary. In: Proceedings Asian symposium Mangrove Environment research management. University of Malaysia and UNESCO, Kuala Lumpur, pp. 226–230
- Lakshmi M, Parani M, Senthilkumar P, Parida A (2002) Molecular phylogeny of mangroves VIII, analysis of mitochondrial DNA variation for species identification and relationships in Indian mangrove Rhizophoraceae. Wetland Ecology and Management 10: 355-362.
- Land M, Grane´li W, Grimvall A, Hoffmann CC, Mitsch WJ, Tonderski KS, Verhoeven JTA (2016) How effective are created or restored freshwater wetlands for nitrogen and phosphorus removal? A systematic review. Environmental Evidence 5: 9.
- Lavieren HV, Spalding M, Alongi DM, Kainuma M, Clüsener-Godt M, Adeel Z (2012) Securing the Future of Mangroves. United Nations University Institute for Water, Environment and Health, Canada. pp.52.
- Lee SY, Hamilton S, Barbier EB, Primavera J, Lewis III RR (2019) Better restoration policies are needed to conserve mangrove ecosystems. Nature Ecology & Evolution 3:870– 872.
- Lewis RR, Brown B (2014) Ecological Mangrove Rehabilitation—A Field Manual for Practitioners. Version 3. Mangrove Action Project Indonesia, Blue Forests, Canadian International Development Agency, and OXFAM. 275 p.
- Lewis RR, Milbrandt EC, Brown B, Krauss KW, Rovai AS, Beever JW, Flynn LL (2016) Stress in mangrove forests: early detection and pre-emptive rehabilitation are essential for future successful worldwide mangrove forest management. Marine Pollution Bulletin 109: 764–771.
- Li J, et al. (2016) Pronounced genetic differentiation and recent secondary contact in the mangrove tree *Lumnitzera racemosa* revealed by population genomic analyses. Scientific Reports 6:29486.
- Lira-Medeiros CF (2010) Epigenetic variation in mangrove plants occurring in contrasting natural environment. PLoS ONE 5: e10326.
- Lo EYY (2003) Phylogenetic relationships and natural hybridization in the mangrove genus Rhizophora from the Indo-West Pacific region. [M.Sc. Thesis]. University of Hong Kong, Hong Kong, 198 p.
- Lo EYY (2010) Testing hybridization hypotheses and evaluating the evolutionary potential of hybrids in mangrove plant species. Journal of Evolutionary Biology 23: 2249-226.
- Lovelock CE, Cahoon DR, Friess DA, Guntenspergen GR, Krauss KW, Reef R, Rogers K, et al. (2015) The vulnerability of Indo-Pacific mangrove forests to sea-level rise. Nature 526: 559–563.
- Ludt WB, Rocha LA (2014) Shifting seas: the impacts of Pleistocene sea-level fluctuations on the evolution of tropical marine taxa. Journal of Biogeography 42(1): 25–38.
- Lugo AE (1986) Mangrove understory: an expensive luxury? J Trop Ecol 2:287–288.
- Lyu H, He Z, Wu C-I, Shi S. 2018. Convergent adaptive evolution in marginal environments: unloading transposable elements as a common strategy among mangrove genomes. New Phytologist 217: 428– 438.
- Mabberley DJ (2008) Mabberley's Plant-Book, 3rd edition. Cambridge University Press, United Kingdom, 1021 pp.
- Macnae W, Kalk M (1962) The ecology of the mangrove swamps of Inhaca Island, Mozambique. J Ecol 50:19–34.
- Magris RA, Pressey RL, Weeks R, Ban NC (2014) Integrating connectivity and climate change into marine conservation planning. Biological Conservation 170: 207–221.

References

- Maguire TL, Saenger P, Baverstock P, Henry R (2000) Microsatellite analysis of genetic structure in the mangrove species Avicennia marina (Forsk.) Vierh. (Avicenniaceae). Mol Ecol 9:1853–1862.
- Majumdar NC, Banerjee LK (1985) A new species of *Heritiera* (Sterculiaceae) from Orissa. Bulletin of Botanical Survey of India 27(1-4):150–151.
- Mall LP, Singh VP, Garge A, Pathak SM, Dandoria MS (1985) A new approach towards the mangrove forest flora of Andaman Islands. Indian Forester 111:290–300.
- Mandal RN, Naskar KR (2008) Diversity and classification of Indian mangroves: a review. Trop Ecol 49(2):131–146.
- McLeod E, Salm R, Green A, Almany J (2009) Designing marine protected area networks to address the impacts of climate change. Front. Ecol. Environ. 7: 362–370.
- Meeuwen MSK (1970) A revision of four genera of the tribe Leguminosae-Caesalpinioideae-Cynometreae in Indo-malesia and the Pacific. Blumea 18: 1–52.
- Mishra PK, Sahu JR, Upadhyay VP (2003) Species diversity in Bhitarkanika Mangove ecosystem in Orissa, India. Lyonia 8:73–87.
- Mohandas M, Lekshmy S, Radhakrishnan T (2014) Kerala Mangroves- Pastures of Estuaries – Their Present Status and Challenges. International Journal of Science and Research 3: 2804–2809.
- Mudailarm CR, Kamath HS (1954) Backwater flora of the West Coast of South India. J Bombay Nat Hist Soc 52:69–82.
- Nabi A, Rao B (2012) Analysis of mangrove vegetation of Machilipatnam coastal region, Krishna district, Andhra Pradesh. International Journal of Environmental Sciences 2: 1754-1764.
- Nabi A, Rao B, Aama Prasad AV (2012) Analysis of mangrove vegetation of Diviseema region, Krishna district, Andhra Pradesh. International Journal of Plant, Animal and Environmental Sciences. 2: 99-108.
- Naskar KR (2004) Manual of Indian Mangroves. Daya Publishing House, New Delhi, India, 220 pp.
- Naskar KR, Mandal RN (1999) Ecology and Biodiversity of Indian Mangroves. Daya Publishing House, New Delhi, India.
- Nasser AKV, Kunhikoya VA, Aboobaker PM (1999) Mangrove ecosystem if Minicoy Island, Lakshaweep. In Marine fisheries information service. CMFRI, ICAR, Cochin, India.
- Navalkar BS (1951) Succession of Mangrove Vegetation in Bombay and Salsette Islands. J. Bombay Nat. Hist. Soc. 50: 157-160.
- Nayak VN, Andrade LV (2008) Floral diversity and distribution of mangroves in the Kali Estuary, Karwar, West Coast of India. National Workshop on 'Mangroves in India: Biodiversity, Protection and Environmental Services'. Institute of Wood Science and Technology, Bangalore. Abstracts, p. 8.
- Nayak VN, Andrade LV (2008) Floral diversity and distribution of mangroves in the Kali Estuary, Karwar, West Coast of India. National Workshop on 'Mangroves in India: Biodiversity, Protection and Environmental Services'. Institute of Wood Science and Technology, Bangalore. Abstracts, p. 8.
- Nehru P, Balasubramanian P (2018) Mangrove species diversity and composition in the successional habitats of Nicobar Islands, India: A post-tsunami and subsidence scenario. Forest Ecology and Management, 427:70–77.
- Ng WL, Chan HT, Szmidt AE (2013) Molecular identification of natural mangrove hybrids of *Rhizophora* in Peninsular Malaysia. Tree Genetics & Genomes 9: 1-10.
- Ng WL. et al. (2015) Closely related and sympatric but not all the same: genetic variation of Indo-West Pacific Rhizophora mangroves across the Malay Peninsula. Conserv Genet 16: 137–150.
- Oliver CD, Larson BC (1996) Forest stand dynamics, Update edition. Wiley, New York.
- Orsini L, Vanoverbeke J, Swillen I, Mergeay J, Meester L (2013) Drivers of population genetic differentiation in the wild: isolation by dispersal limitation, isolation by adaptation and isolation by colonization. Molecular Ecol 22: 5983–5999.
- Osland MJ, Feher LC, López-Portillo J, Day RH, et al (2018) Mangrove forests in a rapidly changing world: Global change impacts and conservation opportunities along the Gulf of Mexico coast. Estuarine, Coastal and Shelf Science 214:120–140.
- Osland MJ, Grace JB, Guntenspergen GR, Thome KM, Carr JA, Feher LC (2019) Climatic Controls on the Distribution of Foundation Plant Species in Coastal Wetlands of the Conterminous United States: Knowledge Gaps and Emerging Research Needs. Estuaries and Coast, https://doi.org/10.1007/s12237-019-00640-z.
- Panda SP, Subudhi H, Patra HK (2013) Mangrove forest of river estuaries of Odisha, India. Int J. Biodivers. Conserv. 5(2): 446–454.
- Pandey CN, Pandey R (2010) The status of mangroves in Gujarat. In Towards Conservation and Management of Mangrove Ecosystems in India/ed. by Bhatt J. R., Macintosh D. J., Ayar T. S., Pandey C. N. and B. P. Nilaratna, IUCN India.141-153.
- Pandey CN, Pandey R (2011) The status of mangroves in Gujarat. In (Bhatt, J.R., Macintosh, D.J., Ayar, T.S., Pandey, C.N., Nilaratna, B.P., eds) Towards Conservation and Management of Mangrove Ecosystems in India. IUCN, India. p. 141–153.
- Parani M, Rao CS, Mathan N, Anuratha CS, Narayanan KK, Parida A (1997) Molecular phylogeny of mangroves III: Parentage analysis of a *Rhizophora* hybrid using random amplified polymorphic DNA and restriction fragment length polymorphism markers. Aquatic Botany 58: 165-172.
- PDO-ICZMP (2004) Areas with special status in the coastal zone. Programme development office for integrated coastal zone management plan project, Water resources planning organization, Ministry of water resources, Bangladesh. Dhaka, Bangladesh.
- Polidoro BA, Carpente KE, Collins L, Duke NC, Ellison AM, et al. (2010) The loss of species: mangrove extinction risk and geographic areas of global concern. PLoS One 5:1–10.

References

- Saenger P (2002) Mangrove Ecology, Silviculture and Conservation. Kluwer Academic Publishers, the Netherlands. p. 360.
- Saenger P, Hegerl E, Davie JD (1983) Global Status of Mangrove Ecosystems; International Union for Conservation of Nature and Natural Resources: Gland, Switzerland.
- Saenger P, Ragavan P, Sheue CR, López-Portillo J, Yong JWH, Mageswaran T (2019) Mangrove Biogeography of the Indo-Pacific. In Sabkha Ecosystems Volume VI: Asia/Pacific, Gul B, Böer B, Khan, MA, Cluesener-Godt M, Hameed A (Eds.) Chapter 23. 379-400.
- Sahu SC, Suresh HS, Murthy IK, Ravindranath NH (2015) Mangrove Area Assessment in India: Implications of Loss of Mangroves. J. Earth Sci. Clim. Change 6(5): 280.
- Sahu SK, Singh R and Kathiresan K (2015). Deciphering the taxonomical controversies of Rhizophora hybrids using AFLP, plastid and nuclear markers. Aquatic Botany 125: 48-56.
- Saintilan N, Wilson N, Rogers K, Rajkaran A, Krauss KW (2014) Mangrove expansion and salt marsh decline at mangrove poleward limits. Glob Chang Biol 20:147–157.
- Sanjappa M (1992) Legumes of India. Bishen Singh Mahendra Pal Singh, Dehra Dun, 338 pp.
- Sanjappa M, Venu P, Arisdasan W, Dinesh-Albertson W (2011) A Review of mangrove species in India. In: Bhatt JR, Macintosh DJ, Nayar TS, Pandey CN, Nilaratna BP (Eds) Towards Conservation and Management of Mangrove Ecosystems in India. IUCN India, New Delhi, 35–50.
- Sanyal P, Mandal RN, Ghosh D, Naskar KR (1998) Studies on the mangrove patch at Subarnarekha river mouth of Orissa state. J Inter Des 2:140–149.
- Saravanan KR, Ilangovan K, Khan AB (2008) Floristic and macro faunal diversity of Pondicherry mangroves, South India. Tropical Ecology 49: 91-94.
- Schmiegelow JMM, Gianesella SMF (2014) Absence of zonation in a mangrove forest in southeastern Brazil. Braz J Oceanogr 62(2):117–131.
- Selvam V, Ravichandran KK, Gnanappazham L, Navamuniyammal (2003) Assessment of community-based restoration of Pichavaram mangrove wetland using remote sensing data. Curr. Sci., 85: 794–798.
- Shaikh SS, Gokhale MV, Chavan NS (2011) A report on the existence of *Heritiera littoralis* Dryand. on the cost of Maharashtra. The Bioscan 6: 293–295.
- Sharma PP, Sikarwar RLS (2014). Floristics of Diu Island. Paper presented in International day for biological diversity island Biodiversity. Uttar Pradesh State biodiversity Board. pp 151-154.
- Sheue CR, Liu HY, Tsai CC, Yong WH (2010) Comparison of *Ceriops pseudodecanda* sp. nov. (Rhizophoraceae), a new mangrove species in Australasia, with related species. Botanical Studies 51: 237-248.
- Sheue CR, Liu HY, Yong JWH (2003) *Kandelia obovata* (Rhizophoraceae), a new mangrove species from Eastern Asia. Taxon 52: 287-294.
- Simard M, et al. (2019) Mangrove canopy height globally related to precipitation, temperature and cyclone frequency. Nature Geoscience 12: 40-45.
- Sindhu SS (1963) Studies on Mangroves. Proc. Indian Acad Sci. 33(8): 129-136.
- Sindumathi S, Vidyasagaran L, Gopakumar S (2014) Phytosociological and edaphic attributes of mangroves in Chettuwai backwater system, Thrissur, Kerala. International Journal of Environmental Sciences. 5. 282-290.
- Satyanarayan Y (1958) Ecological Studies of Bombay Coastline Vegetation. J. Biol. L. Sci. 1: 53-55.
- Shah GL (1962) The Vegetation along the seashore in Salsette Island. Bombay. Bull. Bot. Surv. India 4: 239-240.
- Shaikh SS, Gokhale MV, Chavan NS (2011) A report on the existence of Heritiera lirroralis Dryand. on the cost of Maharshtra. The Bioscan 6(2): 293-295.
- Sindumathi S, Vidyasagaran L, Gopakumar S (2014) Phytosociological and edaphic attributes of mangroves in Chettuwai backwater system, Thrissur, Kerala. Internatinal Journal of Environmental Sciences 5(2): 282-290.
- Singh NP, Prasanna PV, Kulkarni BG (1999) Karnataka. In: Mudgal, V. and P.K. Hajra (eds), Floristic Diversity and Conservation Strategies in India. Vol. 2: 975-996. Botanical Survey of India, Kolkata.
- Sulochanan B (2010) Mangrove ecosystem and its impact on Fisheries. CMFRI Research Centre, Mangalore Karnataka.
- Singh HS (1994) Status report on mangroves in Gujarat State. Gujarat Forest Department, Gandhinagar.
- Singh HS (2003) Marine protected areas in India: coastal wetland conservation. Indian Forester 129(11): 1313–1321.
- Singh VP, Mall LP, George A, Pathak SM (1987) A new record of some mangrove species from Andaman Islands and their distribution. Indian Forester 113:214–217.
- Sippo JZ, Lovelock CE, Santos IR, et al. (2018). Mangrove mortality in a changing climate: An overview. Estuarine, Coastal and Shelf Science. doi: https://doi.org/10.1016/j.ecss.2018.10.011.
- Sippo JZ, Maher DT, Tait DR, Holloway C, Santos IR (2016) Are mangroves drivers or buffers of coastal acidification? Insights from alkalinity and dissolved inorganic carbon export estimates across a latitudinal transect. Global Biogeochemical Cycle, 30 (5): 753-766.
- Smith TJ III (1992) Forest structure. In: Robertson AI, Alongi DM (eds) Tropical mangrove ecosystems. American Geophysical Union, Washington, DC, pp 101–136.
- Snedaker SC (1982) Mangrove species zonation: why? In: Sen DN, Rajpurohit KS (eds) Contributions to the ecology of halophytes, Tasks for vegetation science, 2. Dr. W. Junk, The Hague, pp 111–125.
- Spalding M, Kainuma M, Collins L (2010) World Atlas of mangroves. ISME publication. pp 320.
- Sreelekshmi S, Bijoy Nandan S, Kaimal Sreejith V, Harikrishnan M (2020a) Floristic Structure, Diversity and Edaphic Attributes of Mangroves of the Andaman Islands, India. Thalassas: An International Journal of Marine Sciences. Doi: 10.1007/s41208-020-00191-2.
- Sreelekshmi S, Bijoy Nandan S, Sreejith Kaimal V, Radhakrishnan CK, Suresh VR (2020b) Mangrove species diversity, stand structure and zonation pattern in relation to environmental factors - A case study at Sundarban delta, east coast of India. Regional Studies in Marine Science. Doi: 10.1016/j.rsma.2020.101111.

References

- Sreelekshmi S, Preethy CM, Varghese R, Joseph P, Asha CV, Bijoy Nandan S, Radhakrishnan CK (2018) Diversity, stand structure, and zonation pattern of mangroves in southwest coast of India, Journal of Asia-Pacific Biodiversity, 11: 573-582.
- Swain PK, Rama Rao N (2008) Floral diversity and vegetation ecology of mangrove wetlands in the states of Goa, Karnataka and Andhra Pradesh, India. National Workshop on 'Mangroves in India: Biodiversity, Protection and Environmental Services' Institute of Wood Science and Technology, Bangalore. Abstracts, p. 5.
- Swain PK, Rama Rao N, Mohan S (2008) New mangrove habitats and additions to the flora of Srikakulam district, Andhra Pradesh, India. Indian J. Forest. 31: 431–434.
- Takayama K, Tamura M, Tateishi Y, Webb EL, Kajita T (2013) Strong genetic structure over the American continents and transoceanic dispersal in the mangrove genus Rhizophora (Rhizophoraceae) revealed by broad-scale nuclear and chloroplast DNA analysis. Am. J. Bot. 100: 1191–1201.
- Tansley AG, Fritsch FE (1905) Sketches of vegetation at home and abroad. I. The flora of the Ceylon littoral. New Phytol 4(2–3): 27–55.
- Thom BG (1967) Mangrove ecology and deltaic geomorphology: Tabasco, Mexico. J Ecol 55:301–343.
- Thom BG, Wright LD, Coleman JM (1975) Mangrove ecology and deltaic estuarine geomorphology: Cambridge Gulf-Ord river, Western Australia. J Ecol 63:203–232.
- Thothathri K (1962) Contribution to the flora of Andaman and Nicobar Islands. Bull Bot Surv India 4: 281-296.
- Tomlinson PB (1986) The botany of mangroves. Cambridge (Cambridge University Press). p. 419.
- Tomlinson PB (2016) The Botany of Mangroves, 2nd ed.; Cambridge University Press: Cambridge, UK.
- Thomas KJ (1962) A Survey on vegetation of Veli, Trivandrum with special references to ecological factors. J Indian Bot Soc 42:104–131.
- Troup RS (1921) The silivicultutre of Indian trees, 3 Vol. The Clarendon Press, Oxford, p 1195.
- Twilley RR. et al. (2017) in Mangrove Ecosystems: A Global Biogeographic Perspective (eds Rovera-Monroy, V. H. et al.) Ch. 5 (Springer, Basel, 2017).
- Upadhyay VP, Mishra PK (2014) An Ecological Analysis of Mangroves Ecosystem of Odisha on the Eastern Coast of India. Proc Indian Natl Sci Acad 80:647–661.
- Vasudevan N (2017) Safeguarding the sentinels: the story of mangrove conservation in Maharashtra, Mangal-Van. Mangrove Society of India, Goa, pp. 55–65.
- Venu P, Aridason W, Magesh CR, Satyananda Murthy T (2006) *Brownlowia tersa* (L.) Kosterm. (Tiliaceae) in India. Rheedea 16: 111–114.
- Vidyasagaran K, Madhusoodanan VK (2014) Distribution and plant diversity of mangroves in the west coast of Kerala, India. Journal of Biodiversity and Environmental Sciences. 4: 38-45.
- Vidyasagaran K, Ranjan MV, Maneeshkumar M, Praseeda TP (2011) Phytosociological analysis of mangroves at Kannur Districts, Kerala. Internatinal Journal of Environmental Sciences. 2: 683-689.
- Voris HK (2000) Maps of Pleistocene sea levels in south-east Asia: shorelines, river systems and time durations. J Biogeogr 27:1153–1167.
- Untawale AG (1986) India mangrove ecosystem. In: Technical Report of the UNDP/UNESCO Research and Training Pilot Programme on Mangrove Ecosystem in Asia and the Pacific. Mangroves of Asia and the Pacific: Status and Management. Manila: Natural Resources Management Centre, 467–470.
- Untawale AG, Dwivedi SN, Singbal SYS (1973) Ecology of Mangroves in Mandovi and Zuari Estuaries and the interconnecting Cumbarja Canal of Goa. Indian J. Marine Sci. 2: 47-53.
- Untawale AG, Wafer S, Jagtap TG (1982) Application of Remote Sensing techniques to study the distribution of Mangroves along the estuaries of Goa. In: Wetlands: Ecol. and Management. Proc. 1st Inter. Wetlands Conference, New Delhi. pp. 51-57.
- Van Rheede HA (1678–1693) Hortus Indicus Malabaricus, vol 12. Sumptibus Johannis van Someren et Joannis van Dyck, Amsterdam.
- Vidyasagaran K, Madhusoodanan VK (2014) Distribution and plant diversity of mangroves in the west coast of Kerala, India. Journal of Biodiversity and Environmental Sciences 4(5): 38-45.
- Wang BS, Liang SC, Zhang WY, Zan QJ (2003) Mangrove flora of the world. Acta Bot Sin 45 (3):644–665.
- Wang PX, Wang LJ, Bian YH, Jian ZM (1995) Late Quaternary paleoceanography of the South China Sea: surface circulation and carbonate cycles. Mar Geol 127: 145–165.
- Wang WQ, Wang M (2007) The Mangroves of China. Science Press, Beijing [Chinese].
- Wee AKS, Mori GM, Lira GF et al. 2018. The integration and application of genomic information in mangrove conservation. Conservation Biology. 33:206–209.
- Wee AKS, Takayama K, Asakawa T, Thompson B, Sungkaew S, Tung NX et al (2014) Oceanic currents, not land masses, maintain the genetic structure of the mangrove Rhizophora mucronata Lam. (Rhizophoraceae) in Southeast Asia. J Biogeogr 41:954–964.
- Wee AKS, Takayama K, Chua JL, et al (2015) Genetic differentiation and phylogeography of partially sympatric species complex Rhizophora mucronata Lam. and R. stylosa Griff. using SSR markers. BMC Evolutionary Biology, vol.15: 57.DOI: 10.1186/s12862-015-0331-3.
- West RC (1956) Mangrove swamps of the Pacific coast of Columbia. Ann Assoc Am Geogr 46:98–121.
- Woodroffe CD, Rogers K, McKee KL, Lovelock CE, Mendelssohn IA, Saintilan N (2016) Mangrove sedimentation and response to relative sea-level rise. Annu. Rev. Mar. Sci. 8: 243–266.
- Worthington T, Spalding M (2019) Mangrove restoration potential: A global map highlighting a critical opportunity. 34pp.
- Wyrtki K (1961) Scientific results of marine investigations of the South China Sea and the Gulf of Thailand 1959–1961. NAGA report 2, 164–169.
- Xie W, Zhong G, Li X, Guo Z, Shi S (2020) Hybridization with natives augments the threats of introduced species in Sonneratia mangroves. Aquatic Botany 160: 103166.
- Yeragi SS, Yeragi SG (2014) Status, Biodiversity and distribution of mangroves in South Konkan, Sindudurg Districts, Maharashtra state India An overview. 2 (1): 67-69.
- Zhang R, Liu T, Wu W, Li Y, Chao L, Huang L, Huang Y, Shi S, Zhou R (2013) Molecular evidence for natural hybridization in the mangrove fern genus *Acrostichum*. BMC Plant Biology 13:74-83.

References

- Pool DJ, Snedaker SC, Lugo AE (1977) Structure of mangrove forests in Florida, Puerto Rico, Mexico and Costa Rica. Biotropica 9:195–212.
- Pradhan SD, Palei NC, Rath BP, Swain KK, Kar S (2016) A new addition of mangrove species *Acanthus ebracteatus* Vahl. (Acanthaceae) to the mangrove flora of Bhitarkanika National Park, Odisha, India. Int J Innov Appl Res 4(4):28–31.
- Ragavan P (2015) Taxonomy of mangroves of the Andaman and Nicobar Islands with special references to natural references to natural hybrids of genus Rhizophora. Ph.D Thesis, Podicherry University, pp 293.
- Ragavan P, Dubey SK, Dagar JC, Mohan PM, Ravichandran K, Jayaraj RSC, Rana TS (2019a) Current Understanding of the Mangrove Forests of India. In Dagar JC, et al. (eds.), Research Developments in Saline Agriculture. Chapter 8. 257-304.
- Ragavan P, Dubey SK, Panda M, Roy M, Ravichandran K, Trivedi RK, Jayaraj RSC, Mohan PM, Rana TS (2019b) Critical note on the identity and distribution of Sonneratia griffithii Kurz (Lythraceae) in India- a Critically Endangered Mangrove species. Nordic Journal of Botany. doi: 10.1111/njb.02119.
- Ragavan P, Jayaraj RSC, Saxena A, Mohan PM, Ravichandran K (2015d) Taxonomical Identity of *Rhizophora ×annamalayana* Kathir and *Rhizophora ×lamarckii* Montrouz (Rhizophoraceae) in the Andaman and Nicobar Islands, India. Taiwania 60: 183-193.
- Ragavan P, Mohan PM, Jayaraj RSC, Ravichandran K, Saravanan S (2015c) *Rhizophora mucronata* var. *alokii* – a new variety of mangrove species from the Andaman and Nicobar Islands, India (Rhizophoraceae). PhytoKeys 52: 95–103.
- Ragavan P, Rana TS, Ravichandran K, Jayaraj RSC, Sivakumar K, Saxena A, Mohan PM (2017) Note on identity and distribution of Cynometra iripa Kostel. and C. ramiflora L. (Fabaceae) in the Andaman and Nicobar Islands, India. Check List 13(6): 805–812.
- Ragavan P, Ravichandran K, Jayaraj RSC, Mohan PM, Saxena A, Saravanan S, Vijayaraghavan A (2014) New records of Sonneratia spp. from Andaman and Nicobar Islands, India. Biodiversitas 15: 12-18.
- Ragavan P, Saxena A, Jayaraj RSC, Mohan PM, Ravichandran K, Saravanan S, Vijayaraghavan A (2016) A review of the mangrove floristics of India. Taiwania 61: 224-242.
- Ragavan P, Saxena A, Jayaraj RSC, Ravichandran K, Mohan PM, Saxena M (2015b) Taxonomy and distribution of little-known species of the genus *Xylocarpus* (Meliaceae) in the Andaman and Nicobar Islands, India. Botanica Marina 58(5): 415-422.
- Ragavan P, Saxena A, Jayaraj RSC, Ravichandran K, Saravanan S (2015e) *Rhizophora ×mohanii*: A putative hybrid between *Rhizophora mucronata* and *Rhizophora stylosa* from mangroves of the Andaman and Nicobar Islands, India. ISME/GLOMIS electronic journal 13: 3-7.
- Ragavan P, Saxena A, Mohan PM, Jayaraj RSC, Ravichandran K (2015a) Taxonomy and distribution of species of the genus *Acanthus* (Acanthaceae) in mangroves of the Andaman and Nicobar Islands, India. Biodiversitas 16: 225-237.
- Ragavan P, Saxena A, Mohan PM, Ravichandran K (2013) A hybrid of *Acrostichum* from Andaman and Nicobar Islands, India. ISME/GLOMIS electronic journal 12:9-14.
- Ram AT, Shaji CS (2013) Diversity and distribution of mangroves Kumnalam island of Kerala, India. Journal of Environmental Science, Toxicology and Food technology. 4: 18-16.
- Ram AT, Shaji CS (2013) Diversity and distribution of mangroves Kumnalam island of Kerala, India. Journal of Environmental Science, Toxicology and Food technology 4(4): 18-16.
- Ramachandran KK, Mohanan CN (1987) Perspectives in management of mangroves of Kerala with special references to Komarakom mangroves a bird sanctuary. In: Proceedings national seminar estuarine management, pp 252–257.
- Ramachandran KK, Mohanan CN, Balasubramonian G, Kurten L, Thomas L (1986) The mangrove ecosystem of Kerala; its mapping, inventory and some environmental aspects. Centre for Earth Science Studies, Trivandrum, p 38.
- Ramachandran, K.K. and C.N. Mohanan 1990. The mangrove ecosystem of Kerala, its mapping, inventory and some environmental aspects. Final Report. Centre for Earth Science Studies, Thiruvananthapuram, pp. 100.
- Ramakrishnan R, Gladston Y, Kumar NL, Rajput P, Mani Murali R, Rajawat AS (2020) Impact of 2004 co-seismic coastal uplift on the mangrove cover along the North Andaman Islands. Regional Environmental Change 20:6.
- Ramasubramanian R, Ravishankar T (2004) Mangrove forest restoration in Andhra Pradesh, M. S. Swaminathan Research Foundation, Chennai, p. 26.
- Rao RS (1985) Flora of Goa, Diu, Daman, Dadra and Nagar Haveli. Vol. 1. Ranunculaceae – Caprifoliaceae. Botanical Survey of India, Howrah.
- Rao RS (1986) Flora of Goa, Diu, Daman, Dadra and Nagarhaveli. Vol. 2. Rubiaceae –Selaginellaceae. Botanical Survey of India, Howrah.
- Rao TA, Sastry ARK (1974) An Ecological approach towards classification of Coastal vegetation in India, II. Estuarine border vegetation. Indian For 100:438–452.
- Rao TA, Suresh PV (1990) Ecosystem of Karnataka Coast II: Mangroves. My Forest 26: 399-442.
- Reef R, Feller IC, Lovelock CE (2010) Nutrition of mangroves. Tree Physiol. 30: 1148–1160.
- Rheede. H. Van. 1678-1693. Hortus Indicus Malabaricus. 12 vols. Amsterdam.
- Ricklefs RE, Latham RE (1993) Global patterns of diversity in mangrove floras. In: Ricklefs RE, Schluter D (eds) Species diversity in ecological communities: historical and geographical perspectives. University of Chicago Press, Chicago, pp 215–229..

Fruits and seeds of ...nthus ebrcteatus

References

- Rilov G, Mazaris AD, Stelzenmuller V et al. 2019. Adaptive marine conservation planning in the face of climate change: What can we learn from physiological, ecological and genetic studies? Global Ecology and Conservation 17: e00566.
- Rosentreter JA, Maher DT, Erler DV, Murray RH, Eyre BD (2018) Methane emissions partially offset "blue carbon" burial in mangroves. Sci Adv 4: eaao4985.
- Saenger P (2002) Mangrove Ecology, Silviculture and Conservation. Kluwer Academic Publishers, the Netherlands. p. 360.
- Saenger P, Hegerl E, Davie JD (1983) Global Status of Mangrove Ecosystems; International Union for Conservation of Nature and Natural Resources: Gland, Switzerland.
- Saenger P, Ragavan P, Sheue CR, López-Portillo J, Yong JWH, Mageswaran T (2019) Mangrove Biogeography of the Indo-Pacific. In Sabkha Ecosystems Volume VI: Asia/Pacific, Gul B, Böer B, Khan, MA, Cluesener-Godt M, Hameed A (Eds.) Chapter 23. 379-400.
- Sahu SC, Suresh HS, Murthy IK, Ravindranath NH (2015) Mangrove Area Assessment in India: Implications of Loss of Mangroves. J. Earth Sci. Clim. Change 6(5): 280.
- Sahu SK, Singh R and Kathiresan K (2015). Deciphering the taxonomical controversies of Rhizophora hybrids using AFLP, plastid and nuclear markers. Aquatic Botany 125: 48-56.
- Saintilan N, Wilson N, Rogers K, Rajkaran A, Krauss KW (2014) Mangrove expansion and salt marsh decline at mangrove poleward limits. Glob Chang Biol 20:147–157.
- Sanjappa M (1992) Legumes of India. Bishen Singh Mahendra Pal Singh, Dehra Dun, 338 pp.
- Sanjappa M, Venu P, Arisdasan W, Dinesh-Albertson W (2011) A Review of mangrove species in India. In: Bhatt JR, Macintosh DJ, Nayar TS, Pandey CN, Nilaratna BP (Eds) Towards Conservation and Management of Mangrove Ecosystems in India. IUCN India, New Delhi, 35–50.
- Sanyal P, Mandal RN, Ghosh D, Naskar KR (1998) Studies on the mangrove patch at Subarnarekha river mouth of Orissa state. J Inter Des 2:140–149.
- Saravanan KR, Ilangovan K, Khan AB (2008) Floristic and macro faunal diversity of Pondicherry mangroves, South India. Tropical Ecology 49: 91-94.
- Satyanarayan Y (1958) Ecological Studies of Bombay Coastline Vegetation. J. Biol. L. Sci. 1: 53-55.
- Schmiegelow JMM, Gianesella SMF (2014) Absence of zonation in a mangrove forest in southeastern Brazil. Braz J Oceanogr 62(2):117–131.
- Selvam V, Ravichandran KK, Gnanappazham L, Navamuniyammal (2003) Assessment of community-based restoration of Pichavaram mangrove wetland using remote sensing data. Curr. Sci., 85: 794–798.
- Shah GL (1962) The Vegetation along the seashore in Salsette Island. Bombay. Bull. Bot. Surv. India 4: 239-240.
- Shaikh SS, Gokhale MV, Chavan NS (2011) A report on the existence of *Heritiera littoralis* Dryand. on the cost of Maharashtra. The Bioscan 6: 293–295.
- Shaikh SS, Gokhale MV, Chavan NS (2011) A report on the existence of Heritiera lirroralis Dryand. on the cost of Maharshtra. The Bioscan 6(2): 293-295.
- Sharma PP, Sikarwar RLS (2014). Floristics of Diu Island. Paper presented in International day for biological diversity island Biodiversity. Uttar Pradesh State biodiversity Board. pp 151-154.
- Sheue CR, Liu HY, Tsai CC, Yong WH (2010) Comparison of *Ceriops pseudodecanda* sp. nov. (Rhizophoraceae), a new mangrove species in Australasia, with related species. Botanical Studies 51: 237-248.
- Sheue CR, Liu HY, Yong JWH (2003) *Kandelia obovata* (Rhizophoraceae), a new mangrove species from Eastern Asia. Taxon 52: 287-294.
- Simard M, et al. (2019) Mangrove canopy height globally related to precipitation, temperature and cyclone frequency. Nature Geoscience 12: 40-45.
- Sindhu SS (1963) Studies on Mangroves. Proc. Indian Acad Sci. 33(8): 129-136.
- Sindumathi S, Vidyasagaran L, Gopakumar S (2014) Phytosociological and edaphic attributes of mangroves in Chettuwai backwater system, Thrissur, Kerala. International Journal of Environmental Sciences. 5. 282-290.
- Sindumathi S, Vidyasagaran L, Gopakumar S (2014) Phytosociological and edaphic attributes of mangroves in Chettuwai backwater system, Thrissur, Kerala. Internatinal Journal of Environmental Sciences 5(2): 282-290.
- Singh HS (1994) Status report on mangroves in Gujarat State. Gujarat Forest Department, Gandhinagar.
- Singh HS (2003) Marine protected areas in India: coastal wetland conservation. Indian Forester 129(11): 1313-1321.
- Singh NP, Prasanna PV, Kulkarni BG (1999) Karnataka. In: Mudgal, V. and P.K. Hajra (eds), Floristic Diversity and Conservation Strategies in India. Vol. 2: 975-996. Botanical Survey of India, Kolkata.
- Singh VP, Mall LP, George A, Pathak SM (1987) A new record of some mangrove species from Andaman Islands and their distribution. Indian Forester 113:214–217.
- Sippo JZ, Lovelock CE, Santos IR, et al. (2018). Mangrove mortality in a changing climate: An overview. Estuarine, Coastal and Shelf Science. doi: https://doi.org/10.1016/j.ecss.2018.10.011.
- Sippo JZ, Maher DT, Tait DR, Holloway C, Santos IR (2016) Are mangroves drivers or buffers of coastal acidification? Insights from alkalinity and dissolved inorganic carbon export estimates across a latitudinal transect. Global Biogeochemical Cycle, 30 (5): 753-766.
- Smith TJ III (1992) Forest structure. In: Robertson AI, Alongi DM (eds) Tropical mangrove ecosystems. American Geophysical Union, Washington, DC, pp 101–136.
- Snedaker SC (1982) Mangrove species zonation: why? In: Sen DN, Rajpurohit KS (eds) Contributions to the ecology of halophytes, Tasks for vegetation science, 2. Dr. W. Junk, The Hague, pp 111–125.

Bracnchlets of *Sonneratia lanceolata*

References

- Spalding M, Kainuma M, Collins L (2010) World Atlas of mangroves. ISME publication. pp 320.
- Sreelekshmi S, Bijoy Nandan S, Kaimal Sreejith V, Harikrishnan M (2020a) Floristic Structure, Diversity and Edaphic Attributes of Mangroves of the Andaman Islands, India. Thalassas: An International Journal of Marine Sciences. Doi: 10.1007/s41208-020-00191-2.
- Sreelekshmi S, Bijoy Nandan S, Sreejith Kaimal V, Radhakrishnan CK, Suresh VR (2020b) Mangrove species diversity, stand structure and zonation pattern in relation to environmental factors – A case study at Sundarban delta, east coast of India. Regional Studies in Marine Science. Doi: 10.1016/j.rsma.2020.101111.
- Sreelekshmi S, Preethy CM, Varghese R, Joseph P, Asha CV, Bijoy Nandan S, Radhakrishnan CK (2018) Diversity, stand structure, and zonation pattern of mangroves in southwest coast of India, Journal of Asia-Pacific Biodiversity, 11: 573-582.
- Sulochanan B (2010) Mangrove ecosystem and its impact on Fisheries. CMFRI Research Centre, Mangalore Karnataka.
- Swain PK, Rama Rao N (2008) Floral diversity and vegetation ecology of mangrove wetlands in the states of Goa, Karnataka and Andhra Pradesh, India. National Workshop on 'Mangroves in India: Biodiversity, Protection and Environmental Services' Institute of Wood Science and Technology, Bangalore. Abstracts, p. 5.
- Swain PK, Rama Rao N, Mohan S (2008) New mangrove habitats and additions to the flora of Srikakulam district, Andhra Pradesh, India. Indian J. Forest. 31: 431–434.
- Takayama K, Tamura M, Tateishi Y, Webb EL, Kajita T (2013) Strong genetic structure over the American continents and transoceanic dispersal in the mangrove genus Rhizophora (Rhizophoraceae) revealed by broad-scale nuclear and chloroplast DNA analysis. Am. J. Bot. 100: 1191–1201.
- Tansley AG, Fritsch FE (1905) Sketches of vegetation at home and abroad. I. The flora of the Ceylon littoral. New Phytol 4(2–3): 27–55.
- Thom BG (1967) Mangrove ecology and deltaic geomorphology: Tabasco, Mexico. J Ecol 55:301–343.
- Thom BG, Wright LD, Coleman JM (1975) Mangrove ecology and deltaic estuarine geomorphology: Cambridge Gulf-Ord river, Western Australia. J Ecol 63:203–232.
- Thomas KJ (1962) A Survey on vegetation of Veli, Trivandrum with special references to ecological factors. J Indian Bot Soc 42:104–131.
- Thothathri K (1962) Contribution to the flora of Andaman and Nicobar Islands. Bull Bot Surv India 4: 281-296.
- Tomlinson PB (1986) The botany of mangroves. Cambridge (Cambridge University Press). p. 419.
- Tomlinson PB (2016) The Botany of Mangroves, 2nd ed.; Cambridge University Press: Cambridge, UK.
- Troup RS (1921) The silivicultutre of Indian trees, 3 Vol. The Clarendon Press, Oxford, p 1195.
- Twilley RR. et al. (2017) in Mangrove Ecosystems: A Global Biogeographic Perspective (eds Rovera-Monroy, V. H. et al.) Ch. 5 (Springer, Basel, 2017).
- Untawale AG (1986) India mangrove ecosystem. In: Technical Report of the UNDP/UNESCO Research and Training Pilot Programme on Mangrove Ecosystem in Asia and the Pacific. Mangroves of Asia and the Pacific: Status and Management. Manila: Natural Resources Management Centre, 467–470.
- Untawale AG, Dwivedi SN, Singbal SYS (1973) Ecology of Mangroves in Mandovi and Zuari Estuaries and the interconnecting Cumbarja Canal of Goa. Indian J. Marine Sci. 2: 47-53.
- Untawale AG, Wafer S, Jagtap TG (1982) Application of Remote Sensing techniques to study the distribution of Mangroves along the estuaries of Goa. In: Wetlands: Ecol. and Management. Proc. 1st Inter. Wetlands Conference, New Delhi. pp. 51-57.
- Upadhyay VP, Mishra PK (2014) An Ecological Analysis of Mangroves Ecosystem of Odisha on the Eastern Coast of India. Proc Indian Natl Sci Acad 80:647–661.
- Van Rheede HA (1678–1693) Hortus Indicus Malabaricus, vol 12. Sumptibus Johannis van Someren et Joannis van Dyck, Amsterdam.
- Vasudevan N (2017) Safeguarding the sentinels: the story of mangrove conservation in Maharashtra, Mangal-Van. Mangrove Society of India, Goa, pp. 55–65.
- Venu P, Aridason W, Magesh CR, Satyananda Murthy T (2006) *Brownlowia tersa* (L.) Kosterm. (Tiliaceae) in India. Rheedea 16: 111–114.
- Vidyasagaran K, Madhusoodanan VK (2014) Distribution and plant diversity of mangroves in the west coast of Kerala, India. Journal of Biodiversity and Environmental Sciences 4(5): 38-45.
- Vidyasagaran K, Madhusoodanan VK (2014) Distribution and plant diversity of mangroves in the west coast of Kerala, India. Journal of Biodiversity and Environmental Sciences. 4: 38-45.
- Vidyasagaran K, Ranjan MV, Maneeshkumar M, Praseeda TP (2011) Phytosociological analysis of mangroves at Kannur Districts, Kerala. Internatinal Journal of Environmental Sciences. 2: 683-689.
- Voris HK (2000) Maps of Pleistocene sea levels in south-east Asia: shorelines, river systems and time durations. J Biogeogr 27:1153–1167.
- Wang BS, Liang SC, Zhang WY, Zan QJ (2003) Mangrove flora of the world. Acta Bot Sin 45 (3):644–665.
- Wang PX, Wang LJ, Bian YH, Jian ZM (1995) Late Quaternary paleoceanography of the South China Sea: surface circulation and carbonate cycles. Mar Geol 127: 145–165.
- Wang WQ, Wang M (2007) The Mangroves of China. Science Press, Beijing [Chinese].
- Wee AKS, Mori GM, Lira GF et al. 2018. The integration and application of genomic information in mangrove conservation. Conservation Biology. 33:206–209.
- Wee AKS, Takayama K, Asakawa T, Thompson B, Sungkaew S, Tung NX et al (2014) Oceanic currents, not land masses, maintain the genetic structure of the mangrove Rhizophora mucronata Lam. (Rhizophoraceae) in Southeast Asia. J Biogeogr 41:954–964.

References

- Wee AKS, Takayama K, Chua JL, et al (2015) Genetic differentiation and phylogeography of partially sympatric species complex Rhizophora mucronata Lam. and R. stylosa Griff. using SSR markers. BMC Evolutionary Biology, vol.15: 57.DOI: 10.1186/s12862-015-0331-3.
- West RC (1956) Mangrove swamps of the Pacific coast of Columbia. Ann Assoc Am Geogr 46:98–121.
- Woodroffe CD, Rogers K, McKee KL, Lovelock CE, Mendelssohn IA, Saintilan N (2016) Mangrove sedimentation and response to relative sea-level rise. Annu. Rev. Mar. Sci. 8: 243–266.
- Worthington T, Spalding M (2019) Mangrove restoration potential: A global map highlighting a critical opportunity. 34pp.
- Wyrtki K (1961) Scientific results of marine investigations of the South China Sea and the Gulf of Thailand 1959–1961. NAGA report 2, 164–169.
- Xie W, Zhong G, Li X, Guo Z, Shi S (2020) Hybridization with natives augments the threats of introduced species in Sonneratia mangroves. Aquatic Botany 160: 103166.
- Yeragi SS, Yeragi SG (2014) Status, Biodiversity and distribution of mangroves in South Konkan, Sindudurg Districts, Maharashtra state India An overview. 2 (1): 67-69.
- Zhang R, Liu T, Wu W, Li Y, Chao L, Huang L, Huang Y, Shi S, Zhou R (2013) Molecular evidence for natural hybridization in the mangrove fern genus *Acrostichum*. BMC Plant Biology 13:74-83.

****